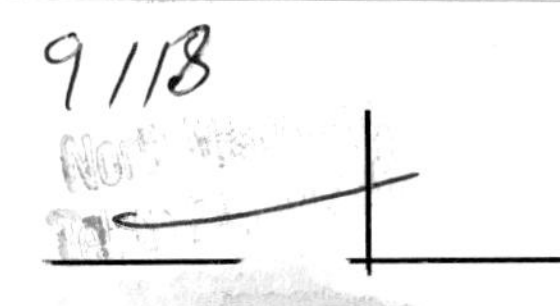

GROSS ANATOMY

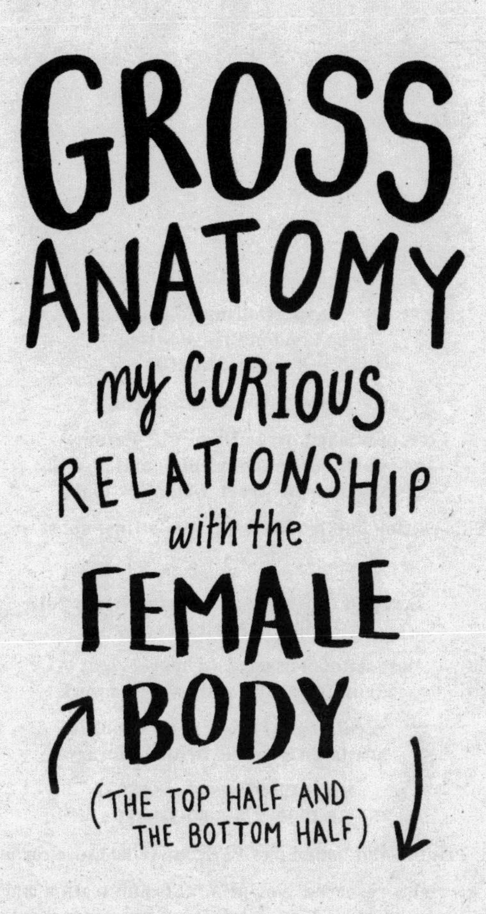

MARA ALTMAN

HarperCollins*Publishers*

HarperCollins*Publishers*
1 London Bridge Street
London SE1 9GF

www.harpercollins.co.uk

First published in the US by G.P. Putnam's Sons
an imprint of Penguin Random House LLC
375 Hudson Street, New York 10014

This edition published by HarperCollins*Publishers* 2018

1

A catalogue record of this book is
available from the British Library

HB ISBN 978-0-00-829270-6
EB ISBN 978-0-00-829271-3

Printed and bound by CPI Group (UK) Ltd, Croydon

For my mom

Body of Contents

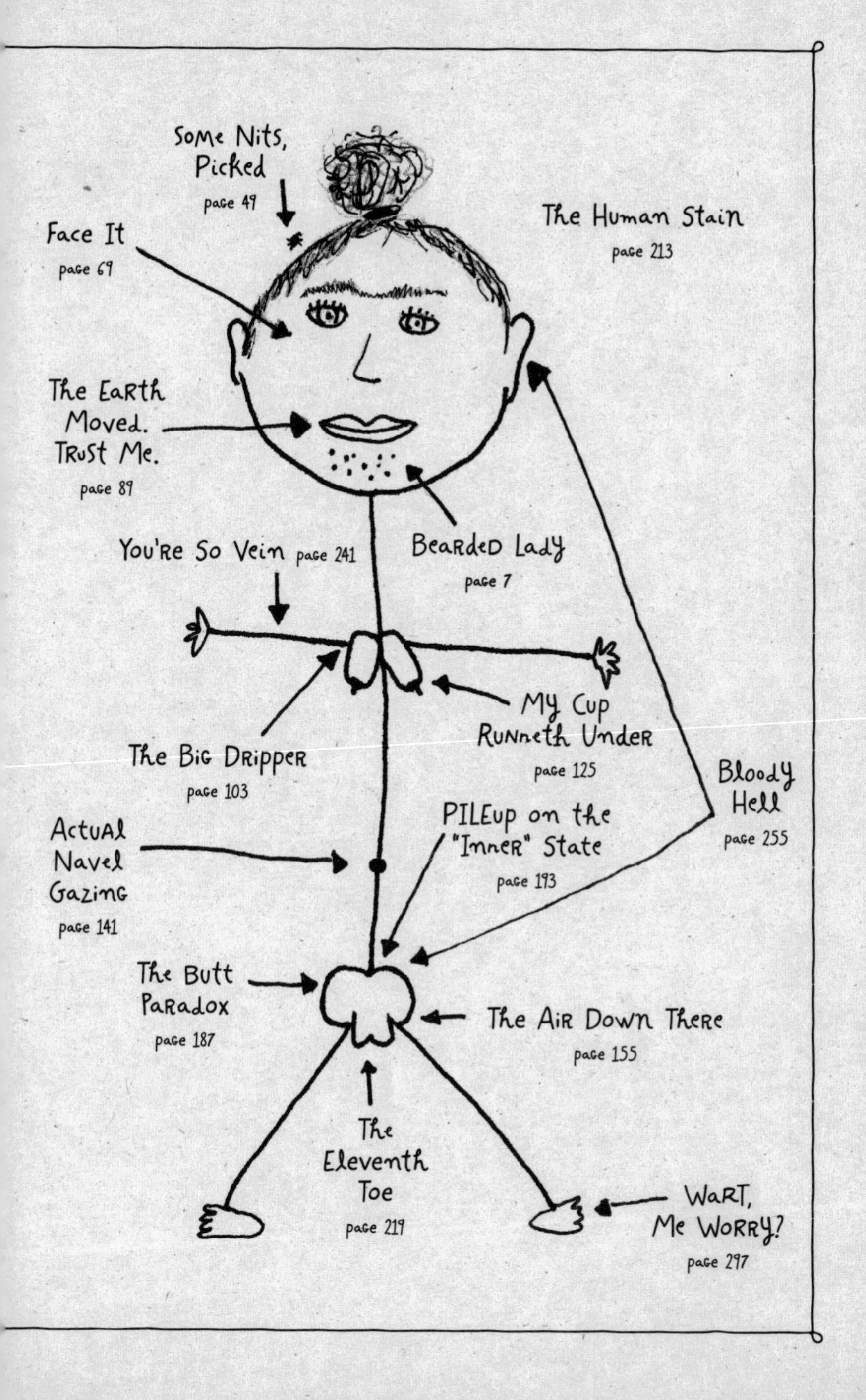

Prologue

To become a master at any one thing, it is said that one must practice it for 10,000 hours. I have been living in my body for 306,600 hours, yet I still feel like a novice at operating this bag of meat. As soon as I feel like I've got everything figured out, something changes—boobs spring out of my chest, I sprout a mustache, floaters homestead in my eyeballs—and I'm left shocked, bewildered, and yet ultimately quite curious. I cannot tell you the number of times I've wondered, especially after a spicy meal, why evolution wasn't smart enough to build us with buttholes made out of something more durable. Lead piping, perhaps?

I'd like to say that I spend my time trying to cure cancer, eradicate hunger, and put an end to global warming, but my brain is naturally inclined toward questions about the human female body. I spend most days wondering about the potential aerodynamic advantages of camel toes and why, when we are built to sweat, I often find myself hiding in a public restroom, drying off my pit stains to pretend that I don't have glands. Why does my dog, every time I squat down, make a beeline for my crotch? The only other thing she's drawn to with such consistency is the garbage can.

I want to be one of those people who, in the morning, sip an espresso while filling in the *New York Times* crossword puzzle—what a respectable hobby!—but instead I'm busy wondering why, as I hump, I never sound nearly as cool and moany as the porn star Sasha Grey in the film *Asstravaganza 3*. Is there a meet-up group for sex mutes?

Let's, for a moment, suspend the idea of self-accountability and attempt to blame these bodily fixations on my parents. They grew up during the 1960s and were the kind of hippies who were so hippie that they refused to be called hippie. "Hippies were so conformist," my mom has always told me.

My parents first met in high school and then dropped out of UC Berkeley together. They began growing plants—mostly cacti and succulents—in their backyard and then, to make a living, sold them to local grocery stores and via mail-order catalogues.

My mom never wore any image-altering materials—no makeup, deodorant, perfume, push-up bras, or high heels. She has refused antiaging creams and would never dream of fillers. (When she read this, she said, "What are fillers?" Sheesh!) She didn't even shave her legs or armpits, and still doesn't to this day. I thought all that was normal female behavior until late elementary school, when I noticed that other moms didn't have a great black muff under their arms when they waved their children in from the playground. I imagined that astronauts could spot my mom from space. "Houston, we have a problem—there appear to be two errant black holes near San Diego's suburbs."

While I felt proud of her uniqueness, I also felt terrified of being ridiculed because of it. I explained to her that it was perfectly possible to wave at me less zealously while gluing her elbow to her side.

So for a long time, I didn't know a lot of woman things. In my twenties, I thought that women tipped the wax lady to keep her quiet.

My father, meanwhile, turned his nose up at anything he deemed unnatural. He hated perfume and artificial scents of any kind. When I tried a spritz of my friend's bottle of White Musk from the Body Shop, he screwed up his face and rolled the car windows down. When he caught me wearing lipstick, he looked at me like I'd just murdered a giant cuddly panda bear to use its innards as war paint.

Growing up, I had a different concept of femininity. I came to think that artificially enhancing my appearance in any way showed a lack of self-acceptance, that it meant I wasn't strong enough to be who I really was. All the girls out there who were wearing makeup, dyeing their hair, and covering their stink were frauds. I, who stepped forth into the world doused in her artisanal BO, was real. Of course, keeping it real doesn't mean that I didn't often feel uncomfortable. I found myself in a constant battle between self-righteousness and shame. Eventually I learned that one's identity can be complemented, not always concealed, by how one chooses to express oneself superficially.

Ultimately, I matured in an environment that made me hyperaware of our social norms because I was constantly conscious of how I was never managing to meet them. Though I now partake in many of the beauty practices that I grew up shunning, maybe it's because of my upbringing that I always catch myself asking, "But why?"

Then again, I'm not sure I can blame my parents for everything. Their aversion to razors probably doesn't account for why I spent the last couple of days mining modern literature for hemorrhoid references or spent an hour unwinding after a rough week by watching Dr. Pimple Popper's blackhead-extraction videos on YouTube.

In any case, I'm not saying that I've got it together more than any other woman; it is precisely my own volatile and apprehensive relationship with my own body parts, such as my bowels, bunions, belly button, and copious sweat glands, that has compelled me to go forth in search of answers from everyone from the goddess worshippers of Bainbridge Island to the top lice experts in Denmark.

This book won't cure a bad hair day or a yeast infection, or anything else for that matter, but it is my hope that by holding up a magnifying glass to our beliefs, practices, and nipples, this book might serve as a small step toward replacing self-flagellation with awe, shame with pride, and vag odor with, well, vag odor is kind of inevitable. But get this, PMS might actually be a superpower!

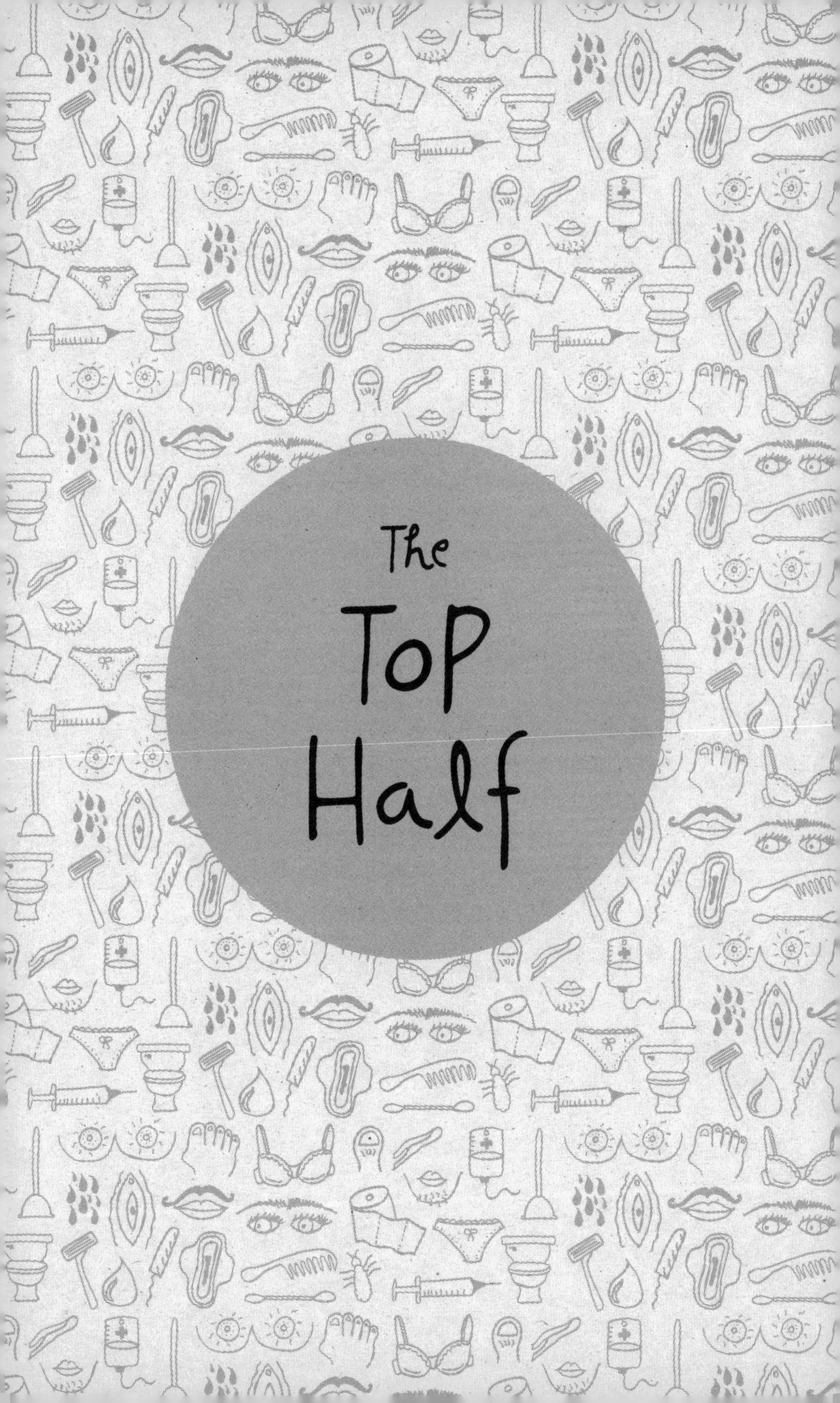
The
Top
Half

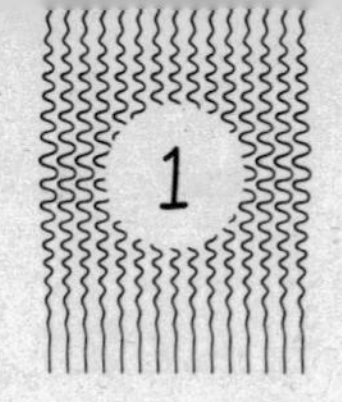

Bearded Lady

It was the turn of the century. I was nineteen years old and a student at UCLA, a school bathed in milky young complexions and spicy Mexican food. I joined friends for dinner at a taco joint on Sepulveda Boulevard, where a dark and deeply handsome young waiter named Gustavo took considerable notice of my face. I will never forget that name, Gustavo. We flirted over the horchata and made googly eyes over the guacamole. My friends evaporated into the atmosphere until it seemed like there were only two of us left in the room. Every time he passed our table, he glanced furtively in my direction, and I returned his interest with the dividend of a smile and the promise of much, much more. It even seemed possible that, at some point in the evening's marathon mating dance, we would speak about more than the Thursday-night specials.

Finally, the check—and our moment—arrived. Gustavo placed the bill in front of my friends and leaned down to my expectant ear. I tingled with excitement about what he might whisper. A phone number ... an address ... a marriage proposal ...

And then they came tumbling from his luscious lips, like poop from a piñata—five simple words that have seared themselves forever into my memory.

"I like your blond mustache," he said.

It is now eleven years later, and I'm on the cusp of marriage to a wonderful man who is covered in hair. He not only makes me feel happy; he also makes me feel smooth. I am writing this story for him, because I have something to tell him.

Dave, I have something to tell you.

I am a bearded lady.

No, not like those women you see at the circus. More like those women you see on the street, in magazines, at the corner coffee shop. Yes, Dave, they're bearded, too. You don't realize it, though, because we are all (except for quite a few Southeast Asians; I'll get to that later) engaged in an endless process of removing the additional and unwanted hair we inexplicably, annoyingly came with.

You see, evolution played a cruel trick on the supposedly fairer sex. It involves chin hair, nipple hair, mustache hair, thigh hair, and—yes—even toe hair. Dave, by God, it's true—we have fucking toe hair! Just like you! But the difference is that we spend millions, no, make that billions, of dollars to have it waxed, lasered, shaved, and otherwise removed from our bodies so that when you see us naked, you won't run screaming into the night.

I'm telling you this now, before we get married, because I am, unfortunately, plagued with two parallel conditions: an inordinate amount of body hair and a genetic predisposition toward brutal honesty. These would seem to be contradictory forces, particularly since I've spent thousands of my own precious dollars in a futile attempt to look as though I'm not a hairy beast. I strapped myself to a wall in Spain and endured the pain of hot wax; I went for monthly laser treatments from a doctor in Bangkok who almost turned my face into a failed lab experiment; I own enough pink disposable razors to affect the quarterly income of Gillette. I've scraped, shaved, yanked, tweezed, and plucked nearly every visible surface of my body, not to mention certain sections I discuss only with my therapist.

I guess I'm telling you this also because I'm trying to figure out why I care. I know you love me no matter what. I realize no one—even you—will ever see the silky brunette strands that occasionally emerge from my nipple. I acknowledge that I'm not the victim of some cruel hormonal joke; I know that plenty of women have it worse than I do.

That raven-haired beauty in front of me at Vinyasa Yoga on Nineteenth Street, Thursdays at four p.m., sports actual muttonchops. But why, when I look in the mirror, do I see Roddy McDowall in *Planet of the Apes*? How can I rid myself of an obsession borne by women since the dawn of time? What weapon do I have to combat the societal standard that all women must be smooth, supple, hairless creatures? When will I be permitted to let my hair down? Not my head hair, but my armpit hair, my facial hair, my leg hair, that little "happy trail." And is that even what I want?

You love me for who I am, right? So why do I want to be somebody else?

I was in my eighth-grade physical education class in suburban San Diego when I learned that there was a really bad kind of body hair to have. And that I had it.

It began with a group of girls, sitting cross-legged on the grass. Our uniforms—maroon drawstring shorts and a gray T-shirt, not that I recall every single solitary detail of that day—revealed our different stages of development. My shirt had ALTMAN written out in black permanent marker just under the peeling, screen-printed figure of our mascot—a crusader. Again, you just kind of remember these things.

While the PE teacher went off to grab soccer balls, we just sat there doing nothing, the sun beating down on us. To pass the time, I was contentedly grabbing one fistful of grass after another and then ripping it out. Grass. Out. Grass. Out. Repeat ad what felt like infinitum.

Finally one of the girls, April, got up and put her hands on her hips. She looked me up and down, but mostly down. She then took a jump back and flung her arms in the air. "Ewww, you don't shave?" April shrieked. "That's SO gross!"

I let go of the grip of grass I had in my hand. The blades of grass fell to the ground, like so many hairs. The girls looked at my legs. I felt like Sissy Spacek at the end of *Carrie*. The hairs sparkled in the sun like beads of blood. Under that withering Southern California sun, they wouldn't stop making a spectacle of themselves.

Other girl legs were splayed around me. It was the dawning of a new era as my eyes scanned them, pair after pair: Shaved. Shaved. Shaved. Shaved. Shaved. Shaved. And then, finally, back to my furry gams, announcing themselves so brightly that they were probably inadvertently transmitting SOS signals to airplanes.

I'd known that women shaved, obviously. At least it had been absorbed by my subconscious. But it wasn't until that moment that I realized I was supposed to join the tradition. I was one of them—a girl—and I had to act accordingly, or be shunned like a leper. My hair apparently represented a possible contagion.

As my fur was inspected by the nearby contingent, a warm rouge attacked the back of my neck and then snuck hotly around to my cheeks. I could pull my legs into my chest and then stretch my shirt over them. I could run away. I could pretend that I didn't hear April and hope that she disappeared. I grabbed another handful of grass and pulled it out, wishing that at that moment each and every one of my leg hairs could be reallocated with such ease.

I was already a little behind. Wait, make that *really* behind. I was roughly a foot shorter than the average eighth grader and had not yet developed a sense of fashion, unless "fashion" could be described as five different colors of sweatpants.

When I was twelve, my mom asked me if I wanted jeans and I declined for practical reasons. "They are too stiff and cold in the morning," I explained. Going shopping was out of the question. I didn't fit into anything in the juniors' section so I had to go to the kids' sizes, where all they offered were variations on flower-print shirts and polka-dotted socks with lace.

Another issue was that I'd practiced gymnastics competitively for the past eight years, and as a result, what had developed were not my breasts, but my thighs. There was a group of guys who, when they spotted me at recess, would shout, "It's muscle girl. Flex!" Those were not the bulges I wanted them to notice.

I couldn't navigate my developmental abyss with conventional tools. So when I got home that day, I dug through the Everything Drawer in the kitchen and found the perfect implement: a battery-operated lint remover. I tucked it into my backpack and went to my room to begin my work.

For some reason I didn't feel like I could ask my mom or dad for a razor. I felt guilty even considering the request. I knew if I did so, I would be knocking their entire modus operandi. They saw the world through their late-1960s Berkeley-colored glasses and maintained a loyalty to All Things Natural—countering societal conventions like hair removal, and maybe having something to do with nostalgia for John Lennon's unkempt eyebrows. Meanwhile, my mom hadn't removed hair on any part of her body, ever.

And my dad professed to love it. "I've been very happy with this hairy little creature," Dad would say.

In addition to his shaving shibboleths, Dad often made the point that he did not like it when women wore makeup or perfume (yes, that includes deodorant). Basically, we were a hair-positive household that practiced a Don't Hide How You Came doctrine. But instead of feeling free to be who I was, sometimes this hairy-go-lucky attitude felt confining. Again, I have to bring up the White Musk. My dad wasn't cool with even a little spritz of White Musk, and who didn't like White Musk?

Apparently, the entire family had met secretly at some point without me and formed a pact against all forms of body enhancements and alterations.

Once, I'd put on some lipstick and my older brother asked, "Why are you wearing that stuff?" The question was so laced with condemnation that I felt like he'd found me shooting up heroin.

I pointed out to him that he was dating a girl who shaved and wore blush and concealer and lipstick and eye shadow and mascara and also some sort of raspberry scent that I felt certain I'd once also whiffed at the Body Shop. He said those weren't the parts he liked about her. But at thirteen I could connect the dots; he was attracted to girls who gussied up. Guys liked girls who gussied up. Still, I couldn't help feeling ashamed that I'd tried to change my innate lip shade in front of my makeup-mocking family.

When my brother went back to his homework, I looked in the mirror and rubbed off the fakery. I wanted to fit in.

But getting rid of my hair wasn't exactly about improving my looks. I didn't quite comprehend what a female leg should look like at that point anyway—and I wasn't trying to attract a guy yet. At thirteen, guys remained as untouchable as tropical fish in an aquarium. I admired their firm fins and bright colors as they passed, but we could never blow bubbles together. They didn't even notice my nose pressed up against the glass.

No, I had to remove hair for basic schoolyard survival, or risk permanent exile to the farthest reaches of the lunch area. I dreaded the idea of being called "gross" again. During that time of major pubescent shifts, April made it her job to strain out confusion—a self-appointed quality-control officer on the San Marcos Junior High School playground, barking at any girl who failed to maintain her proper place on the feminine side of the distinct gender line.

That meant no leg hair, ladies.

For the remaining hours of that school day it had felt like forty million sniper eyes were laser-focused on my legs. Even the slightest pupil flicker bound in my direction caught my attention. The embarrassment was vaguely equivalent to having toilet paper hanging from your shoe, but not really. You can't shake off leg hair. I know; I've tried that, too.

So I locked the door to my bedroom and pulled out the lint-remover contraption. I flipped on the switch. It started buzzing. I lowered it to my calf, feeling equal measures of shame for having hair and for buzzing it off with a machine. I cringed as it made calf contact, expecting excruciating pain. But it really only tickled, asserting itself as a machine manipulated for the wrong purpose.

Hair was not lint. I needed a plan B.

I couldn't steal a razor from my mom, like my girlfriends could from theirs, because she didn't have any. Although my father used blue Bic disposables for his cheeks, the commercials made it quite clear that legs needed pink.

After a week of wearing pants, I finally got the gall to ask my mom about shaving.

"Are you sure you want to do this?"

To leave her ranks, I'd be a traitor. She'd be out a hairy compatriot. Me, her only daughter—her own flesh and blood—straying from the path.

But. I. Couldn't. Not. Do. It.

I nodded.

Mom bought me a disposable pink razor and some shaving cream and accompanied me to the master bathroom. She handed me the equipment and sat on the toilet seat, expectantly, as I planted my foot on the edge of the bathtub.

"Now what?" I asked.

"I don't know," she said. "I guess you just slide it up your leg."

"You think that's all you have to do?"

"Try it," she said.

Clearly, she was clueless.

"Like this?" I said, moving the razor over my shin.

The razor left an empty path in its wake. Look, Ma—no hair!

I could now return to the schoolyard and show April my shiny, glamorous new gams.

THINGS I LEARNED IN KINDERGARTEN

By sophomore year, I was finally getting on track. Much to my pleasure, my pubes had sprouted. I'd look at them in the shower and think, *I made those!* I remained in hair heaven for two entire anxiety-free years. If I'd have known that they'd be the only two years of relative hair peace I'd ever experience, I would have taken time to appreciate them more, maybe made a documentary. I was riding high, experiencing my first boyfriends. I discovered that boy pubes looked a lot like girl pubes.

Did I mention that I had pubes? I had pubes! We all had pubes!

But then, all of a sudden—late in my junior year of high school—an assemblage of keratin and protein had conspired beneath my skin to march out of a large number of tiny holes. And not just holes hidden where no one could see them. They were on my upper lip!

I'd noticed these little hairs on my upper lip before, but I'd ignored them—they were little blond wispy nothings. But now they were getting a little darker and a bit longer. If I caught myself in the right light in my bedroom, I could see a vague resemblance to Tom Selleck.

How in the fucking shitball motherfucking hell did I get a mustache?

Only males had mustaches. I was not a male. Or was I?

I remembered that my mom had this stuff called Jolen, in a small turquoise box with white lettering. When I was younger, I used to watch her work its magic. She would mix some powder with cream. The substance would get fluffy and bubbly—the astringent compound burning our nostrils. She would spread the yogurt-like goop on her upper lip and wait ten or so minutes before washing it off. Underneath the bleach, the hair would get so light that it was practically invisible.

At the time, I wasn't able to see the apparent hypocrisy. If my mom was so liberal and wanted to stay "all natural," then why would she lighten her upper-lip hair?

That was a question I would be able to ask only later.

For now, I took that turquoise box from her cabinet. I decided that news of my mustache would be known only to my closest friends, Shannon and Natasha—one a blond Caucasian and the other Cambodian, both of whom grew very fine and small amounts of hair (and the latter of whom is so hairless that waxers, over the years, have often felt guilty charging her full price for any one service; looking back, I should have had a hairy Italian girlfriend or two).

Shannon and Natasha bleached with me. With the white goop swabbed thickly on our upper lips, we looked like we were starring in a road production of "Got Milk?" We turned it into a ritual. While the bleach did its work—tingling and then slowly building up to a stinging sensation—we turned off all the lights so that my bug-shaped glow-in-the-dark stickers would burn green, and sat in a circle, singing aloud to the Cranberries.

In your head, in your head
Zombie, zombie, zombie-ie-ie . . .

After I washed off the mixture, I felt relieved.

That hair, as far as I was concerned, became invisible. I just had to keep up the ritual every two to three weeks.

I was all set.

Until I met Gustavo.

How many people had noticed my "blond mustache" and didn't tell me? I tried to recall different boyfriends and situations. I'd kissed plenty of boys by then. Had they gotten mustache burn from my face? Is that why Sam didn't ask me on that second date? Or Jonathan? Or Bill? Is that why that cashier at Vons, the grocery store, was looking at me strangely when I bought razors for my legs? How did I not realize that with my olive skin tone, bleaching my hairs until they were practically white might create a situation on my face?

Gustavo was the first man to ever mention my body hair, but I had collected enough data to make me pretty sure that men were, as a gender, opposed to it.

A few years before, I was listening to Adam Carolla and Dr. Drew talking to this complete jerk on the radio show *Loveline*. The caller was complaining about his girlfriend's nipple hair. He said he found it nasty and couldn't get turned on when he saw the little strands. He was thinking of breaking up with her. I was shocked to learn that women got nipple hair—and thrilled to check and discover that I'd been mercifully spared that fate—but now, three years later, as I stood in horror after spotting my very own first nipple hair, I knew I faced certain rejection from any man who encountered this new deformity.

I was beginning to understand that there was a very small window of what was "acceptable" and I had ventured beyond it. It wasn't long after the Gustavo Fiasco that I noticed, while staring down at my bikini area, that my pubic hair had been marching, steadily and without heed, down my legs as if it could practice homesteader rights on the rest of my body.

I was now nineteen years old, and it was time for my first visit to a bikini waxer, whom I came to think of as an aggressive border control agent, getting rid of undocumented pubic immigrants. When she entered with the wax strips, I smiled awkwardly and asked the question that I'd be asking for the rest of my life in any and every hair-removal situation.

"Am I normal?"

She said that I was, but I didn't believe her.

"Are you sure?" I said.

"We've all got hair," she said.

I knew that we all had hair, but that wasn't the question. I wanted to know where exactly I stood on the hairy scale, because that was becoming the problem. Ladies were ripping out their hair before I got a good look at it; therefore I was feeling like a beast among a hairless breed.

She proceeded to rip out the hair that jumped the border—about half an inch—but then she spotted the hair on my stomach. For quite a while, I'd had a light "happy trail" from my belly button downward. It was the inspiration for a nickname—Happy—that I'd acquired at fifteen. For a while, I'd considered the name cute.

"You want me to get that, right?" she said, spreading the wax on it before I answered.

"Why, is that not good?"

Rip.

"Well, you probably want to get rid of it," she said, throwing my happy trail in the trash.

And that's how I learned that apparently happy trails aren't as happy as they sound.

By the age of twenty, I was finally coming to terms with the fact that no hair was considered good hair except for the hair on your head, eyelashes, and eyebrows, and those only if they were in the right shape. Arm hair, it seemed, got a pass as well, even though it didn't look any different from leg hair, which is weird. But even toe hair had to go. I didn't even know that I had toe hair, but then it turned out that I did, which was bad. I'd always remember that I forgot to get rid of it when I'd fold my torso over my legs in yoga, and then I wouldn't be able to stop myself from staring at it.

For the crotch, news of the Brazilian style—going completely bare—that would soon sweep the USA had not yet reached my ears. I still thought it was normal to keep all the pubic hair except for the bits that peeked out from my bathing suit. And though I trimmed a little off the sides every now and again, I was proud to have a bush. And I continued with the normal stuff—shaving, plucking, and waxing. I also fell into a dependent relationship with Sally Hansen home wax strips—prewaxed plastic in a rectangular shape. I just had to rub it between my palms to heat up the wax and then I could rip out my hair myself. The problem was that I had issues with getting all the excess wax off, so by the end of the day, I'd end up with an accumulation of colorful fuzz and lint that made wherever I waxed look like my skin was growing patches of sweatshirt.

When I went to Spain for my year abroad as a college junior, I got my legs waxed while being strapped vertically to a wall with a leather belt. I felt a bit vulnerable, but I didn't question it as long as the wax did its job.

I went to India in 2003, the year I finished undergrad, to work at a newspaper, and got my entire face threaded. I said I wanted only the upper lip and eyebrows done, but Smita just kept going. She touched my cheeks and said, "Face?" I shrugged. She took that as a signal to wind up her thread and tear out all the fuzz from my cheeks, chin, and jowls.

Paid professionals were always trying to get rid of more and more of my hair. It happened again when I went to a bikini-waxing joint in Greenpoint, Brooklyn, a few years later. I just wanted a little off the sides, as the bush had been growing out for quite a while. When the waxer saw me—saw *that* part of me—she looked into my eyes with a fortune-teller's boldness and shook her finger back and forth.

"The man does not like dis," she said. She put her fingers toward her tongue, pretending to pinch out hairs. "Plaaaa plaaaa," she said. Then she got all dramatic and faked a male choking episode. She slathered on the hot wax and said calmly, "Very good dat you are here."

When we were done, she unzipped her pants to show me her bald pussy. "Look at it," she said. "Look. No hair." Then she tried to convince me to sign up for laser. "Plaa plaa," she explained again as she zipped up her pants. "They do not like dat."

The only thing I really came to enjoy about hair removal was the inevitable ingrown. There is nothing—and I mean it, nothing—more fundamentally satisfying than extracting a hair that's been growing in the wrong direction. Period. Call it my nurturing side.

Little did I know the worst was yet to come. What happened next made me yearn for the days when a blond mustache was my only problem.

I was twenty-three. I was about to start a one-year journalism master's program at Columbia and was getting a facial at Mario Badescu Skin Care salon on East Fifty-second Street in Manhattan. Everything was going well until the buxom Russian woman examining my face with a bright light rubbed my chin.

"You zchuld git reed of dis," she said.

How did she see them? I thought I was the only person who knew.

She busted my years of self-denial. Toppled them. Crushed them into tiny shards. It's like when you have a big red volcanic pimple and you just convince yourself that you're making it out to be a much bigger deal than it actually is and most likely no one notices it, but then some friend says, "Ouch, that looks like it must hurt." And they are pointing at your big red volcanic pimple that no one is actually supposed to be able to see, so you say, "What must hurt?" and they say, "Your big red volcanic pimple." And you cover your face with one hand and say, "Oh, you can see that?" And they say, "Well, it *is* a big red volcanic pimple."

So it was true. I had chin hairs that people could actually see. They were real. Like, actually there.

Hairs growing out of my chin!

I mean, I knew about them, of course, but I also didn't. I believe my inability to recognize them as an entity—as a growing, living, real part of my body—stemmed from my self-preservation instinct. I'd even plucked them before, but I'd managed to convince myself immediately afterward that I hadn't. My chin was smooth, dammit!

But now the jig was up. I started scanning my chin every morning for one of those evil hairs to reappear. I began carrying tweezers and a mirror in my purse.

I told no one of this new calamity. At least when I discovered my upper-lip hair, I knew that other women shared my shame. Upper-lip waxes were offered at salons. I'd never seen a chin wax mentioned anywhere, and I didn't want to ask anyone about it, in case they told me they'd never heard of such a heinous thing.

I started having these disturbing fantasies that totally freaked me out: I have a mental break and go to a loony bin, but there's no one there to pluck me. When I envision Insane Mara, I'm more embarrassed about the stray hairs than I am about the fact that I've completely lost my mind and am trying to make love to a trash can.

Or what about when I'm old? Old Mara's hands are going to be so shaky from all the meds and her eyesight will be deficient, so there's no way she's going to be able to pluck with any kind of proficiency.

Or maybe Old Mara has Alzheimer's and her grandkids will come visit as she stares at a wall and thumbs the hem on her shirt. "Is Grandma a he or a she?" they'll say. I'm more embarrassed for Alzheimer Mara's hair than for the fact that she thinks her nephew is her husband.

Or I get run over by a car on some New York street and I'm in a coma. My family rushes to Coma Mara's bedside and they look at one another in shock, not because of my medical status, but because they realize I'm different from what they thought I was. "Oh my!" Mom says. "Did any of you know Mara had a goatee?"

I knew that there were many more important issues going on in the world and that my worrying about such an insignificant bodily matter was selfish and maybe even bordering on narcissistic, but I couldn't help my feelings. I was irrational. Global warming was spawning under my skin. Genocide was happening on my face.

I finally had to talk to someone about it, and it was during my winter break from Columbia that it finally burst forth.

"Mom, I've got chin hair!"

"But I don't see it."

"It's there," I said.

She came in closer.

"Don't come too close!"

"Why not?"

"'Cause then you'll see it!"

She blamed it on my dad's side of the family and never spoke of it again.

I continued to pluck my way through my master's program, and from then on kept my chin hairs to myself. But in the midst of all this, I began dating a guy. We were fooling around—nuzzling, hugging—one day in Central Park. Tenderly, he put his hand on my face. "I love the fuzz on your face," he said. "It's so soft." He then made a downward stroking motion from my cheek all the way to my chin. That moment may have seemed romantic to him, but it was the closest I'd ever come to shitting myself besides that one time I had dysentery and was stuck on a twelve-hour bus ride from Dharamsala to Delhi. I turned in the other direction as quickly as possible and encouraged him to fondle my hoodie.

I would never put myself in that position again:

Natural sunlight.

Bare face.

Man at close range.

After attending grad school, I moved to Bangkok for a job as a features writer at a Thai newspaper.

In retrospect, not the best idea in the world for a hairy Western five-footer with budding self-esteem issues.

Thai people, as it turns out, aren't hairy. They don't have any hair except on their heads. They seemed like magical people to me with all their hairlessness, like they lived in some kind of fairy-tale world. I kept looking for hair, scanning crowds for it to reassure myself that I was normal. Maybe I was overreacting—at this point I'm pretty sure I had some form of body-hair dysmorphic disorder—but I often felt like if I stopped plucking, I'd be able to grow more impressive facial hair than most Thai men. That thought made me feel so unsexy that it's hard to properly explain.

That's when I decided to try "permanent reduction" methods for the first time. It was 2005 when I finally signed up for laser. Once a month, I would go to a Bangkok hospital called, I swear, Bumrungrad. I'd lie on a gurney in a brightly lit room. All blank white walls, slightly yellowed by time. A doctor would come in with gloves, goggles, and a mask on over his face. A nurse would cover my eyes with darkened goggles and swab jelly on my skin. The doctor would then spend about ten minutes zapping my face with something that looked like the suction side of a Hoover. I had to fold my tongue over my upper front teeth so that when they did my upper lip, I wouldn't feel the pain of the laser reaching my gums or whiff the slight smell of melting enamel. After, they'd give me icepacks for my red face, which emitted so much heat that my cheek, if placed on a woman's abdomen, could probably help relieve menstrual cramps.

It couldn't have been very healthy, but I wasn't thinking about that then. I had one goal in mind: complete eradication. I'd ride home on the back of a motorcycle taxi and stay home for the night, until the swelling had receded.

I should have realized that there was a problem. I've always been kind of cheap. For example, I won't pay ten bucks for a sandwich that would give me nutrition and probably pleasure—six is my top price—but I could somehow rationalize spending a thousand dollars for someone to fry my face.

On my last visit, they elevated the laser a bit too high and it burned my upper lip. I still have the scar. It's about the size of a raindrop. When I'm cold, it turns white.

When people ask where I got the scar, I tell them, "One time I was making soup—some sort of bean stew—and it was boiling so wildly that it splattered me. . . . Yeah, just like that, third-degree burn. Crazy, right?"

Yeah, right.

It was embarrassing to admit that I made myself look worse by trying to look better. It still is.

Even right now.

Yep, still embarrassing.

But not only was I embarrassed; I also felt ashamed. I was back to being that kid poised with the lint remover over my leg—feeling equal shame for having hair as for getting rid of it. Why couldn't I just be okay with who I was? Why was I spending so much money and time hiding myself?

But if you thought I'd stop it with the laser after realizing all that, then you haven't been reading this very closely.

Two years later, in the middle of my second laser treatment back in New York, I began to consider the possibility of a medical problem. I felt like I was fighting a rare battle—but I wasn't sure because, theoretically, if other women were like me, it would be a battle fought alone and behind closed doors. If other women were waging it, I wouldn't know. But then again, could any of them have so many wanton whiskers? This couldn't be what was supposed to be happening to a woman's body.

So I went to my gyno for a follicular assessment and possible intervention.

Unfortunately, she had some bad news for me: I was normal. She explained that there are three common reasons for unusual quantities of hair on women. They either have polycystic ovaries or hormone imbalances, or they were simply born with hairy genes. "Many Eastern Europeans have a lot of dark, thick hair," Dr. Chrisomalis said. I could have sworn that she was examining my chin as she spoke.

A waxer once told me that she knows what she's about to deal with before people even take off their pants because the eyebrows reveal everything. Why couldn't my doc just check out my eyes, then?

"But it's got to be something else," I pleaded. I'd recently contemplated the possibility that I'd hit early menopause—there had been some hot flashes, I'm pretty sure—and I'd never given up that early idea that I might be part man. I speculated now that my nuts just hadn't descended yet. "I've got hair even on my . . ."

But I couldn't tell a medical professional about the nipple hair. And what would be the point, anyway? I'd plucked that morning especially for her.

"I don't think you have PCOS," she said. "Other symptoms are weight gain and acne, but if it'd make you feel better, we can do some tests and maybe some blood work on your hormone levels."

She extracted some of my blood and scheduled me for an ultrasound. That actually got me a tiny bit excited. It would be awesome if something was medically wrong. I'd be officially diagnosed and on my way to a cure. I could stop going crazy.

But the ultrasound revealed nothing wrong with my ovaries. No cysts. There weren't even any hidden male gonads. When my gyno got back to me about the blood tests, she said that all my hormone levels were normal.

"Normal? Are you sure?"

"Totally normal."

So my doctor was telling me it's normal to be a hairy beast. I was relieved, terrified, and lost.

I couldn't quit the laser. I continued treatments at a place called American Laser, on Broadway near Twenty-second Street in Manhattan. In the waiting room, they had magazines like *People* and *OK!* in a pile. I think they put them there for a reason; they wanted me to look at Kim Kardashian's poreless and follicle-free face and get turned on about having my body blasted with a machine I didn't understand in the slightest.

I dislike those magazines and think of them as vapid and a waste of time, but that's only because I can get sucked into them for hours and I always end up feeling guilty about my desire to know how many hours a day Angelina leaves her kids with the nanny, instead of using my time to start understanding the crumbling economy.

So I'd get into the laser-treatment room, conjure the hair-free cover girl, and tell the laser lady to put the damn thing on the highest they could without causing my face permanent damage.

"It's going to hurt," she'd say.

"I don't care," I'd say.

"Tell me if it's too high."

"It's not high enough!"

Hair brought out a little bit of psycho in me. I never acted like that anywhere else, except for maybe when I'm baking. (I get really bossy when I'm baking.)

The American Laser office was in the same building as a casting agency. Sometimes on the elevator ride up, I'd pretend to mouth some scenes from *A Streetcar Named Desire* and reapply ChapStick in the mirror so that no one would suspect that I was actually lasering.

No, silly, I'm not hairy. I'm an actress.

I also kept it from the guy, Dave, whom I'd started dating in 2008. I would throw away the laser appointment cards so that he couldn't find them and instead use code—"lunch with Leslie" or just an exclamation point—when I wrote down the appointment in my calendar.

When I moved in with him in 2010, a whole new challenge emerged. Close quarters put my secret in jeopardy. I carried out my depilatory duties like they were a covert Navy SEAL operation. I had extra razors and tweezers in my gym bag and purse and hidden in bathroom corners. Mixed martial arts fights were my saving grace. Dave would be attached to the couch for hours at a time, watching hairless men grapple each other, while my stainless-steel Tweezerman and I got it on in the bathroom. If Dave asked what I was doing in there for so long, I'd tell him I was picking at pimples or that the milk in my coffee was working its way through my intestines. That usually shut him up.

The point is, I'd rather have Dave think I was shitting than plucking. His knowing that I was so hairy would have rendered me faulty, almost broken—like he'd driven off with a lemon from the used-car showroom. But I also yearned for him to know and accept me as I was. I realize it doesn't help our relationship that the only thing I can think about when we cuddle is how to position myself to keep him from seeing any stray hair that might break free. To be perfectly honest, I wouldn't even be writing this if we weren't already engaged. Publicly divulging my hairiness during my dating years would have ruined my ratings on Jdate and Match.com.

You can't sell a car by pointing out the jagged, deep dent on the driver's side.

I hate that I feel that way, but there it is.

And as long as I'm talking about things I hate—this is a little off the point, but you know what always kills me? It kills me when girls compliment my eyebrows, because in the aughts, eyebrows with girth came back into fashion. "Wow, they're so nice and thick. I wish I had those," friends would say. The compliments are always by women who are fair-skinned and light-haired. I've never had a thick-browed lady say one thing about my eyebrows. You know why? Because they know the behind-the-scenes story. If any of those light-haired ladies knew what those two caterpillar-shaped suckers actually meant, they'd back away from the situation with their hands up.

Anyway, I kept up the laser treatments for three years.

After my last appointment, I asked to speak to the office manager.

"It didn't work," I said.

I wanted my chin as hairless as a piece of polished granite or my money back. Even though I knew the truth—that laser can be very good for dark hair (pubic, armpit, man beard), as it targets the melanin in the follicle—it has a much harder time getting rid of fine and lighter hair like the gang of strays I had on my face.

"Well, the face is a very stubborn place," the office manager said. "We always tell all our clients that. If you want, we can sign you up for another treatment."

"Why should I sign up for another treatment when it didn't work after three years?"

"The face is a very stubborn place," she reiterated.

"If it's stubborn, why should I do more laser?"

"It takes time," she said. "The face is stubborn."

I stared at her. Then she giggled.

"Why are you laughing?" I asked.

She straightened her posture and relaxed her mouth.

"This is not funny," I said, raising my voice. "I'm. Still. Hairy!"

I got up and walked out without finishing the conversation. I left that place knowing that I couldn't go back, but kind of wishing I could lock myself in one of their treatment rooms and shoot the laser at my face until the SWAT team came and ejected me.

I knew that I was sick, but I didn't know of any other way to become comfortable with myself besides burning my skin off with a weapon.

So over the months since the doctor's appointment and my last laser session, I was in a hair purgatory, contemplating my next move. Instead of just going moment to moment, working to eradicate each hair as it surfaced (though I did that, too), I began thinking more about an odd irony. To be a complete woman, I felt as though I had to get rid of a part of myself. But why? Why does there have to be all this shame and angst about something that's a natural part of being a woman? The pressure to be hairless has driven me to feel like I have to hide something from my fiancé, to spend thousands of dollars, to feel less worthy than my female peers.

For years I've been pretending that I don't have something that I quite clearly have. That takes a lot of energy.

I like getting answers to questions, so I pretended to be an objective reporter and called up *Allure* magazine. I asked to speak with the beauty editor, Heather Muir. To be honest, I disliked Heather before I even spoke to her. I disliked her because of what she represented, and also because her name conjured the image of downy blond hair on her thighs, the sort one doesn't even have to shave. Also, even if I might follow some beauty customs set forth by magazines like Muir's, I'm generally opposed to people imposing their subjective view on millions of women. It's because of people like Muir that I've put myself through so much hair-removal pain over the past fifteen years that if I experienced it all at once, it would likely be lethal.

"Overall you want to be presenting yourself as really groomed and well kept, and unwanted hair falls in that category," Muir explained. "Maintain and take care of it to look your best and be polished."

Listening to her kind of made me want to strangle myself. "Why do you think we get rid of our hair?" I asked, trying desperately not to slam the phone down.

"We do it to feel better about ourselves," she said. "And so we're more socially accepted."

This chick was definitely blond. I could feel it. Or maybe Cambodian.

Muir used the actress and comedian Mo'Nique, who showed up with hairy legs at the Golden Globes in 2010, as a warning. "It was so taboo and people were embarrassed and laughing," Muir explained. "She's an example of 'Oh my gosh, I never want to be that girl.'"

Muir went on to talk about trends for the bikini, and she quoted Cindy Barshop, who founded and runs the Completely Bare salons, named after Barshop's own initials. The salons specialize in laser hair removal. Barshop was most recently in the news after PETA's condemnation of her fox-fur merkins (also known as pubic wigs). Yes, in a paradoxical move, she wanted you to rip off your own fur and then glue colorful feathers and animal fur to your genitals.

I knew what Muir was talking about. I'd just recently experienced my first Brazilian wax. It was for Dave's birthday in October. I waxed everything off for him, except for a small triangular shape (the formal term, I suppose, would be "landing strip").

He liked it. A lot.

I got upset that he liked it. "What, you don't like it when I'm natural? When I'm me . . . all me?"

"I like that, too," he said. "I like you every way you come."

"It seems like you like this more."

"Weren't you doing it for my birthday because you knew that I'd like it?"

"Yeah, but . . ."

That's when I realized—wait, actually, I realized nothing. I'd endured yet another painful ritual, but for reasons I couldn't explain to my boyfriend or to myself.

Ultimately, it felt strange not having hair there. At one time I had been so proud of the hair and then it was gone and its disappearance appreciated. I didn't feel like I had a vagina anymore; now it was a baby bird—pink and freshly broken out of its shell—that I'd stuffed down my pants and was suffocating between my legs. Besides, I never realized until I was bare how useful the hair had been over the years when I'd find myself in the shower without a loofah. If the muff could do one thing—and it can do more than one thing—it could make a really nice lather.

I thought I was enterprising with my lather trick until I read in *The Naked Woman* by Desmond Morris about a tribe living on the Bismarck Archipelago in the South Pacific who used their pubic hair to wipe off their hands whenever they were dirty or damp. In the same way "as we are accustomed to using towels."

The most horrific thing, though, about the wax was when the pubic hair grew back. It looked like mange, and felt like chicken pox.

So, back to Cindy Barshop, who is basically the Queen of Clean. If *Allure* and other beauty magazines were using her as a source—as much as it made me fear for the future of America and the mental health of all the hairy women who populate it—for fairness' sake I needed to go see this woman at her Fifth Avenue location, to hear her side of the story.

Barshop was on Season 4 of *The Real Housewives of New York City*. That means she is tall and skinny, with a lot of cheekbone and full lips. I had issues with her on principle.

"It's fashion," Barshop said, sitting in the back office of her salon, a corner sectioned off with French doors from the baroque-inspired waiting room. "I mean, we all know it. A woman should have no hair on her face. It should be groomed and nowhere else do you want to see hair. I mean, no one says, 'Oh, okay, let's have hairy arms. That looks great.'"

But I would. I would totally say that!

"Do you ever think it's okay to have a unibrow?" I asked. I did have arm hair, and wanted to steer this supposedly objective interview toward some practical information I could use.

She looked up from her phone; she had been texting as I spoke. "What do you think?"

I thought I wanted to shove Barshop's phone down her throat. Instead I skipped to my next question: "And the bikini?"

"Completely bare," she said. "That's really where it's gone."

"So what does that mean as far as landing strips are concerned?"

"That's so old," she said, laughing.

"How old is that?"

"Must be five to seven years old."

"Oh, I just got one."

Silence.

And in that soundless gap, Barshop had managed to tell me that my vagina was so out of style that it was basically wearing a matching velour hoodie-and-pants set from Juicy Couture.

She then told me about a new hair-removal line that she's coming out with for girls—eleven- to thirteen-year-olds—to safely remove their hair at camp.

At this point in the conversation, I began to fixate on her upper lip. I couldn't stop. It was this perfectly smooth blanket of bare skin. At the same time, I found myself loathing everything she seemed to stand for; I couldn't help coveting her hairlessness. I couldn't see even one strand of fuzz anywhere on her. Did she douche with laser?

I finally asked the malevolent woman if she feels good about what she does. I left out the part of my question that went "... destroying the minds and values of millions of women everywhere."

"I don't really think of that very often," she said.

Finally, an answer that I could believe!

"But yes, because having hair on your face or somewhere else not great is a very emotional thing. If you're uncomfortable, you withdraw. So yeah, I feel good about what we do."

The truth is that I understood what she was talking about. I've felt the same way. But I wondered if she thought our society could ever become hair-friendly enough to eliminate the discomfort.

"I just can't imagine it," she said, stroking her hairless chin. "It's like saying being heavy is better . . . it's the same thing. Like it used to be okay, having an extra twenty pounds was the look, but I don't think we're going to regress back to that. We've evolved."

Barshop, throughout our interview, had continued to look down at her phone and text; she was doing it again. Right now.

"I can tell you want to go," I said, summoning politeness from some deep recess of my rage.

"Oh, you're so sweet," she said.

No, I'm not, Cindy. I actually hate you a little bit.

Cindy was, truly, the nemesis of a woman's ability to choose. She's the type of person who narrows beauty into such a small space that hardly anyone can fit in; she makes us hate ourselves. Now, when I look in the mirror and feel misery about the ugly strays straddling my chin, I realize it's her eyes that I'm looking through.

When I got back to the street, I mumbled angry somethings as I looked down at my arm hair. I was so insecure that one little comment about arm hair could make me question the past thirty years of keeping it. I didn't want to pretend that I didn't give a shit anymore; I wanted to be like my mom and *really* not give a shit. As I mulled that over, I went to run some errands. I ended up at Aveda to grab some shampoo. While there, I noticed some dark hairs—like wiry muttonchops—on the face of the lady helping me, and I was thinking: *See, Cindy Barshop, she can live with it. Right on! You go, lady with cheek hairs! Empowered hairy ladies rule!* Then I went all retroactive on myself and started thinking, *But does she know about them?* Should I tell her about her cheek hairs? She must want to know about those cheek hairs. I mean, she couldn't have actually wanted them there, right?

"You want some tea?" she asked.

Aveda gives you free tea.

"No, no," I said, backing away. "I don't want tea."

I managed to keep my mouth shut.

Even though I have weird hairs, I couldn't help being judgmental about other women's weird hairs.

I realized that it happens all the time. When I see a lady in the street with a mustache—the same mustache I could easily grow (except for that scarred part that doesn't grow hair anymore)—the thoughts in my head are so shitty. It goes from *Right on, you nonconformist powerful woman* to *I'll totally let you borrow ten bucks so that you can take care of that.*

I don't like that my brain does that. I really don't.

I wondered if I was any better than Muir and Barshop.

When I got back home, I realized how incapable I was of realigning my thoughts. I'd have to be hypnotized or brainwashed to think hairy was okay. The revulsion felt so deep-rooted that I couldn't help finding the strands more or less . . . well, yucky.

While people like Muir and Barshop upheld the ideals of hairlessness and maybe even expanded on them (and I would continue to dislike them for that), they didn't invent them. I took the next couple of days to read some books and studies on hair removal. I wanted to know when and why this idea of hairlessness as an ideal first entered our heads.

I got really into it, blitzing those books with my highlighter. I found out that women's hair removal isn't even that old of a practice.

The Europeans were hairy when they came over to America. Hairy colonies. Very hairy colonies. Even up to one hundred years ago, women were letting it all hang naturally.

The hair landscape started changing in the early 1900s, when advertising became national via countrywide-distributed magazines like *Ladies' Home Journal* and *Harper's Bazaar*, which, along with touting Crisco and Kleenex, began promoting clean-shaven pits.

At the same time, women's fashions were also changing. Sleeveless gowns became the rage, and the hemline moved from the ankle up to the mid-calf in 1915, eventually reaching just below the knee in 1927. Women were showing more skin than ever before, which meant they were also showing more hair.

The year 1915 began a period that historian Christine Hope labeled "The Great Underarm Campaign." This is when advertisers got nasty.

About a dozen companies, including that of King C. Gillette—who less than two decades earlier had come out with the first disposable razor—waged nothing less than full-on character assassination against women with underarm hair. Magazine ads used words to change the connotation, referring to the hair as "objectionable," "unsightly," "unwelcome," "dirty," and "embarrassing." On the other hand, hairless women were described as "attractive," "womanly," "sanitary," "clean," "exquisite," "modest," and "feminine."

Kirsten Hansen, in her 2007 Barnard College senior thesis, "Hair or Bare?"—which would have been salve to my teenage angst had I found it in ninth grade; think *The Catcher in the Rye* for disgruntled hairy girls—explained that advertisers tried to relate outward cleanliness with inner character. "Advertisers invoked moral values like modesty and cleanliness that had been central to Victorian America," she wrote, "and linked them to the modern value of exterior beauty."

I found the ads insanely horrible, yet quite psychologically compelling and to the point.

One, in 1922, raised the pertinent question "Can any woman afford to look masculine?" and followed with this answer: "Positively not! And moreover, there is no excuse for your having a single hair where it should not be."

The battle against leg hair came next, in a stage that Hope coined as "Coming to Terms with Leg Hair." Leg-hair removal didn't catch on quite so quickly, mostly because women could cover up their legs with stockings.

The upper class adopted the trends first, as hairlessness had been marketed as a status symbol, but by the 1930s, the practice had trickled down to the middle class. The hairless-leg deal was sealed during World War II, when stockings became scarce.

These ads made me angry, but for some reason, these ads caught on; they must have spoken to something—an insecurity or a lack or a desire—because they stuck so profoundly.

The idea that leg hair is gross is so ingrained that, as one study I read revealed, during puberty twice as many girls as boys develop a fear of spiders. When asked to describe the spiders, girls more often than not depicted them as "nasty, hairy things." This happens around the same time they start getting rid of their own body hair. Spiders! Sheesh.

Why did we embrace hairlessness? When I spoke with Jennifer Scanlon, a women's studies professor at Bowdoin College and the author of

Inarticulate Longings: The Ladies' Home Journal, *Gender, and the Promises of Consumer Culture,* she told me that women shouldn't be seen completely as victims of the advertisers. "Women had a role in this, too," Scanlon said.

That figured.

She explained that women were searching for something at that time; they wanted self-esteem, sensuality, and independence. "The culture wasn't offering them these things," Scanlon explained, "but advertisers did. They said if you remove your armpit hair, you're going to feel like a sensual being."

"So," I said, "you're saying that instead of hair removal, the advertisers could have just as easily been like, 'Chicken livers are the answer. Rub these livers all over your body and you will feel sensual.'"

"Yes," Scanlon said, "it was about filling a need."

But the ads for leg hair and pit removal weren't the worst thing I learned.

When I met up with my friend Maggie one morning for coffee and a discussion of my reporting to date, I told her what I perceived to be the worst. "Dude, ladies irradiated themselves to remove hair!"

"What?!" she said.

I'd found out about this in an article written by Rebecca Herzig, a professor of gender studies at Bates College.

When radiation, and more specifically the X-ray, was discovered in 1896, scientists found that besides killing carcinomas, it also eradicated hair. X-ray epilation clinics opened up all over the United States.

By the early 1920s, there were already reports that exposure to radiation could be dangerous. Yet clinics continued to stay open and offer the hair-removal service. Women were lured in by the idea of a "pain-free" procedure and kept there by brochures espousing everything from social acceptance to the socioeconomic advancement that would come from obtaining "smooth, white, velvety skin." They specifically targeted immigrant women who might feel marginalized because of their foreign (and hairier) origins, which I, a hairy Jew, related to.

Maggie, a hairy Italian, also understood.

By 1940, the procedure was outlawed, so these radiation salons began operating in back alleys, like illegal abortion clinics. Many women suffered gruesome disfigurement, scarring, ulceration, cancer, and death, all because of the extreme pressure to become hairless. The women who were adversely affected were dubbed the "North American Hiroshima Maidens," named after the women who suffered radiation poisoning after the nuclear bombs hit Japan in World War II.

To some women, hairlessness has literally been worth dying for. As depressing as that was, I kind of admired it.

Maggie brought her hands to her mouth and her eyes got big. "That's a monstrosity!" she said. "That's batshit crazy."

"Mags," I said, "I think I would have been one of those chicks. I would have stuck my face right into some radioactivity."

Clearly, I still had some issues.

I continued to call on more academics for information.

Oh, who am I kidding? I was calling them for comfort.

For the past eight years, Bessie Rigakos, a sociology professor at Marian University, has studied why women remove their body hair. Her biggest challenge in finding answers has been that she cannot find enough women who don't remove their body hair to use as a control group in her studies.

Before volunteering for her next study, I began with the basics.

Why do we remove our body hair?

"I research hair removal," she said, "and I do it myself, and I still don't know why we do it, which is amazing."

I felt better already.

She went on to say there are so many factors involved that she just can't pinpoint which exactly is the cause. "I wish I had the answer," she said. "Is society controlling it or are women controlling it?"

Keep going, Bessie. I'm wondering the same thing myself.

One thing Rigakos definitely believes is that hair removal gives women positive feedback and is thus a positive force. "Just like how when kids pee in the potty, they are rewarded," she said, "when women adhere to beauty standards, then they are rewarded in society."

Somehow that analogy lost me, and I hung up from my call with Rigakos just as uncertain as before, but at least I felt academic validation in my uncertainty. Rigakos had a doctoral degree in hair-removal studies from Oxford, or something like that.

Next, I called Breanne Fahs, a professor of women's and gender studies at Arizona State University. Fahs was incredibly passionate on the subject and spoke rapidly. Which was good, because I was getting married in less than three months and needed some quick answers.

"It's amazing how people imagine hair removal is a choice and not a cultural requirement," she said. "If they say it's a choice, I say try not doing it and then tell me what you think."

"What would happen?" I asked.

She said the practice of growing body hair can be so intense that it can show women how marginalizing it is to live as an "Other." Growing hair, she means, will give you a taste of what it's like to be queer, be fat, or have disabilities.

"You experience this tidal wave of negative appraisals of your body," she explained.

"How do you think it came to be this way?" I asked.

"At the root of this is misogyny," she said. "It's a patriarchal culture that doesn't want powerful women. We want frail women who are stripped of their power." She explained that in Western culture, men are fundamentally threatened by women's power and eroticize women who look like little girls. "We don't like women in this culture," she said. "Pubic-hair removal is especially egregious. It's done to transform women into prepubescent girls. We defend it and say it's not about that, that it's about comfort. They say they don't want their partner to go down on them and get a hair stuck between their teeth as if that's the worst thing that could ever happen to them."

When I got off the phone with her, I admit, I felt pretty tense. She made hair removal sound like it was the beginning of the end of this civilization. I didn't need that kind of responsibility.

I needed to know if there were any reasons why, evolutionarily speaking, humans might be more attracted to hairlessness. I have to acknowledge that during my reading, I did find evidence that even though hair removal wasn't popular in early America, it has been done on and off for as long as humans have existed.

Archaeologists believe that humans have removed facial hair since prehistoric times, pushing the edges of two shells or rocks together to tweeze. The ancient Turks may have been the first to remove hair with a chemical, somewhere between 4000 and 3000 BC. They used a substance called *rhusma*, which was made with arsenic trisulfide, quicklime, and starch.

In *Encyclopedia of Hair: A Cultural History*, Victoria Sherrow explains that women in ancient Egypt, Greece, and the Roman Empire removed most body hair, using pumice stones, razors, tweezers, and depilatory creams. Greeks felt pubic hair was "uncivilized"—they sometimes singed it off with a burning lamp. Romans were less likely to put their genitals in such peril and instead used plucking and depilatory creams. When in Rome . . .

This means that though I'd like to place all the blame on advertisers, maybe they were just jumping on an inherently human trait and exploiting it legitimately.

I called Nina Jablonski, an anthropology professor at Penn State, to find out why, from her human evolution–informed perspective, women might be viewed as more attractive when they are hairless.

"Things that are considered to be attractive are also most childlike," she said, "and hairlessness is something we associate with youth, children, and naked infants."

She obviously hadn't seen my baby pictures.

Jablonski went on to explain that women who are considered attractive often have facial attributes that exaggerate youthfulness and are reminiscent of children—thinner jaw, longer forehead, big eyes relative to the rest of the face, plump lips, small nose, and shorter distance between mouth and chin.

"In MRI studies, a huge part of the brain indicates affection, love, and an outpouring of positive emotion when a person lays eyes on a child," she said. "So these same responses could be elicited in a man when he sees a woman with childlike attributes."

Interesting, I thought—but I didn't particularly like to hear it. I was suddenly starting to feel like I might want to embrace my natural state at last, and didn't want evolution to get in the way of what was considered beautiful.

So I asked Jablonski why facial hair on a woman is more taboo than any other hair on the body—taboo to the point that we not only hide it, but hide that we got rid of it. I was hoping that her answer might help me at last divulge my darkest secret to Dave.

First, she assured me that having some facial hair in women was normal.

That was a fabulous and very comforting start to her answer.

She went on to explain that it's because the follicles on men's and women's upper lips are more sensitive to androgen and especially testosterone. She said that "peach fuzz" is seen on the upper lip of a pubescent male as his testosterone ramps up and before the appearance of the larger-diameter hairs of the mustache and beard. Because women also have androgen, though at lower levels than males, peach fuzz develops on their upper lip. "That is the normal state in many mature women," said Jablonski.

So my mustache that I flipped out about as a high school junior was actually a normal symptom of puberty? Sweet! Though a little late.

But wait.

Jablonski wasn't done yet. She went on to list the reasons women might feel compelled to rip off their totally natural upper lipstache.

First, she offered the obvious notion that most women don't want to be mistaken for a pubescent male. "It gives mixed sexual signals," she said.

Mixed?

Second, she said that women, as they get older, have more androgens and fewer estrogens. "Facial hair becomes more visible and less 'peachy' as women age," she said. "And they get even more obsessed with removing it because they want to look ever more youthful."

So basically, I gathered that women with less facial hair appear younger, and since more facial hair is correlated with menopause and therefore a higher age, having less could essentially give signals of continued fertility.

Got that?

And isn't that the driving force of humans and all animals, really? We're all in this, theoretically, to reproduce, right? So maybe, from a strictly academic perspective, I'd been getting rid of my face hairs all this time so that men would see me as a qualified baby maker before I'd even really consciously thought about if I wanted to make babies myself.

Now I was hopelessly confused.

The next day, I was talking to my friend Erin. I was finding that as I researched hair, I was becoming desensitized to the taboo and could speak more freely about my own hair issues, so I ended up telling her about my latest chin hair.

Erin, much to my delight, admitted to having some chin hairs, too. "I discovered one back in high school while I was in math class," she said, bringing her hand to her chin. "I was just thumbing my chin like this and then there was this little thing." She had discussed the hair with two of her friends who also had chin hair, and they had employed one another to be emergency pluckers if one ever fell into a coma or became otherwise incapacitated.

"Seriously?" I said.

I was somewhat astonished, but also pleased to know that I wasn't alone—in having the chin hairs or, even more unexpectedly, in the ongoing fear-of-coma scenario.

Over the next couple of weeks, I interviewed close to twenty women about their body hair, of whom more than a few also had a plan in place for their strays if they ever were not able to pluck on their own. For some, the surrogate plucker was their mother. For others, it was a sister or a friend. So far, I haven't heard of the position being filled by a husband or boyfriend.

It felt good to know that I wasn't alone, but it also bothered me to know that so many of us lived in such fear that our biological side would show. It was bad enough that we occasionally had to be seen in natural sunlight.

So on November 14, I began growing out my body hair. I contemplated growing the chin hairs, too, but I figured that I would probably incur some minor to medium psychological damage as a result. I wasn't substantially practiced in the Zen arts of shrugging off contemptuous remarks.

Even a friend, Ali, warned me, "Don't do it for your own mental health." Ali and I have a lot in common. She's so freaked out about her own hair that her husband doesn't know she Nairs her face and bleaches her arms.

Her biggest fear is that when she has a baby, her husband will see her breastfeeding in daylight. "He'll see my boobs and they are going to be so sore, so I don't know if I'll be able to pluck," she said, "and does it bother your child if there are weird hairs there?"

Meanwhile, nothing really dramatic occurred as my hair grew in. It was sparser than I'd expected. My legs were not particularly hirsute, popping up with fine dark hair about a quarter- to a half-inch long. They looked the way a wood floor at a salon would look after a stylist had trimmed a balding man. The armpits, however, came in fuller. They developed a brown fuzz, which was surprisingly soft. Sometimes when I'd reach my arms upward, I'd think I'd spotted something—like a rodent—in the periphery, but then when I'd swing my head back to look, I'd remember that it had actually been my new armpit locks.

I felt some anxiety about going to yoga and the gym—where my legs and underarms were on display—wondering what people were going to think of me. But mostly I felt like a rebel. I wanted someone to say something and I wanted to defend my choice, but no one even seemed to look in my direction.

Only once did I see two girls laugh and point at my armpits. I was self-conscious about it, but I also felt a little relieved. All these years of hair angst haven't been for nothing. People actually can be judgmental schmucks!

The absolute coolest thing—and it wasn't actually that cool—was when I stood naked in front of a full-length mirror with my arms raised and noticed that, with the hair under my arms, it looked like I had two decoy vaginas. I suspected that, somehow, those were used to much advantage during our cavewoman days.

The empowerment that I'd hoped would come, though . . . it just didn't.

A lot of the time I just felt hairy, and everything was a little worse for it:

The dishes are dirty . . . and I'm hairy.

Something is rotten in the fridge . . . and I'm hairy.

I have no money . . . and I'm hairy.

I felt like my body was morphing outside its jurisdiction—crisp lines were suddenly blurring. I was a coloring book and a little kid was coloring outside the markings. My eyebrows broke free from their usual shape and simultaneously were trying to visit my hairline and my nose. How did Frida do it?

To feel momentary relief, I'd visit the Hairtostay.com website, which called itself "The World's ONLY Magazine for Lovers of Natural, Hairy Women." It was part female-hair-fetish porn site and part positive hair treatise. You can do everything from have hairy phone sex to peruse articles such as the one titled "Are Hairy Legs a Deterrent to Crime?" It wasn't to commiserate with other hairy women that I went there, though. I went to stare at ladies who were hairier than I was so that I could feel smooth for a change.

It was finally December—time for my family's annual vacation together. This year we were going to Southeast Asia, land of the genetically hairless women. Right before we left, I bought a box of Sally Hansen prewaxed strips (that addiction had never evaporated) and ripped off my happy trail. That was the one hairy part of my anatomy that I just couldn't take anymore. And once it was torn off, I actually felt like I could breathe deeper.

I was soon in Cambodia with my family. When we went to Angkor Wat, a temple complex from the twelfth century, I asked my tour guide, Vutta, how Cambodians felt about women and body hair.

"They don't do anything to the hair," he said. "Well, actually, they don't really have the hair."

"So no waxing or shaving?"

"Actually, the girls want to have light skin like you."

"But if they get light skin, they will have the hair that comes along with it."

"To be honest," Vutta said, "the people here believe that a girl with the hair is lucky. She can get a better life. A better husband."

"Really?" I said. That was the most hair-positive belief I'd heard, probably ever.

"But it's not true," he said. "They just believe it. We are so behind in our economy and society because people believe silly superstitions like that."

"So it's not lucky to have hair?"

"Not any more lucky than not having hair."

"Oh."

At this point, I began to think I was actually journeying backward.

On the final day, I got one of my legs threaded on the beach in Vietnam. I did it as an experiment. I'd never had threading on anything except my face before. Besides, the woman who did the threading had been chasing me for the past three days, pinching my hairy legs as I passed.

I sat down on a little platform that she had propped up in the sand, about five feet from the water. I was shielded from the sun by a big umbrella. The hair, by this time, was about a half-inch long. The woman wound the thread around her hand and put one part of the loop in her mouth. She twisted the thread and then bent down and started ripping out my hair. It felt like a pack of mice were sinking their jaws into my skin over and over again. I grabbed at the sheet covering the platform below me. I felt the sweat slide down my arm as I yelled "Ouch!" again and again and again.

She leaned over me, and each time I said "Ouch," she said, "No ouch later, later beautiful."

I was amazed that the same hairless aesthetic prevailed on the other side of the world.

I quit after half of one leg. I couldn't handle the pain. A razor seemed so much more humane. I was also having trouble letting go of the hair. I hadn't come to an understanding with my body hair yet. That is, I still didn't really like it. I felt guilty for favoring my leg without the hair, being so thrilled with how smooth it looked—that is, until I sat down and spoke to my mother. I'd been putting it off, but it was time, since it was our last day of the trip. She would be going back to California, and I would be heading back to New York.

My mom and dad were sitting on wooden chaise lounges on the beach. Mom was in sunglasses, a hat, and a bathing suit, comfortably showing off her legs and pits. They weren't as intense as I remembered them. I don't think an astronaut would be able to see them from space, which is how I used to feel when she'd pick me up after school, waving for me to come over with her tank top on.

I sat down beside her, crossing my hairless leg under the hairy other one. "So, were you guys bummed when I started shaving?"

"I wasn't that happy about it," said my dad. "Natural is better, but it's your business. I just thought it might be a problem for you later, get you on the wrong track."

"Which track?" I asked.

"Well, you cut your hair and they branch and then you cut it again and they branch."

"Are you thinking about pruning trees?" I said.

"Yeah," he said, "that's how I see it."

I'd always assumed that my mom didn't shave because of her radical self-acceptance—and I yearned to be like that, to accept myself in my all-natural state—but we never really had a conversation about it before, and here she elaborated.

"I got into the politics," she said. "I also read a lot of Zen and Buddhist texts, and it really felt like accepting who I was was more important to me than looking a certain way for society."

As she said that, something clicked for me that hadn't before. The Jolen!

"Well, if you're so Zen and comfortable with yourself, then why do you wax your upper lip hair?" Her Jolen bleach habit, by this time, had turned into a wax habit.

She paused to think about it for a moment. She started and then stopped. Then started again. "I guess you're right," she said. "I wax my upper lip, and I think my face looks better when I do. It's probably that it worked into my cosmetic feeling about myself, so I guess I can't claim to be this Zen person who would flaunt all."

I'm pretty sure it was at that moment that my perspective began to shift, but I wouldn't realize it until I was back in New York. For the moment, I just thought it completely coincidental that on the evening I had that conversation with my mom, alone in my hotel room, I decided to shave off all the hair I'd grown for the past two months.

Weeks after we got back from Southeast Asia, I was sitting on the sofa with Dave in our East Village apartment. I hadn't done laser for nine months. I'd just finished writing the 14,000 or so words you just read. I put a sofa pillow in my lap and inched toward the corner of the couch. I stared at him until he looked away from a *Law & Order: Special Victims Unit* rerun, the one where some guy has a fetish for recording people urinating in public bathrooms and accidentally witnesses a pedophilic sex crime.

Maybe I could have waited for better timing.

Or maybe, maybe, it was the perfect time.

"What?" Dave said, noticing that I was focused on him, not on Detective Stabler's interrogation.

"I want you to know that I have chin hairs," I said.

He smiled slightly, cocked his head to the side, and returned his focus to the fetishist.

"I'm serious. I do."

Dave looked over at me now, searching his mind for the appropriate thing to say, but I didn't give him a chance to respond.

I told him in rapid-fire narrative the whole story of my hair fixation as fast as the man in the old Micro Machines commercials—the doctor, the laser, the morning pluckings, the purse tweezers, and how when he looked at me in a certain way, I feared that he wasn't actually looking at *me*, that he was searching for errant follicles on my face.

Slowly, Dave began to lean forward. Closer. And closer. Still closer.

"What?" I pleaded. "What?"

Dave didn't say anything. Suddenly he was only inches away; he could see every pore on my face, every hair on my body. His big, soft brown eyes loomed over me like microscopes.

I wiggled in fear of being found out.

Then he slapped me lightly on the cheek. "Get it together," Dave said. "It's just hair."

Good point.

We leaned into each other, arms and lives forever intertwined, and turned back to the television set.

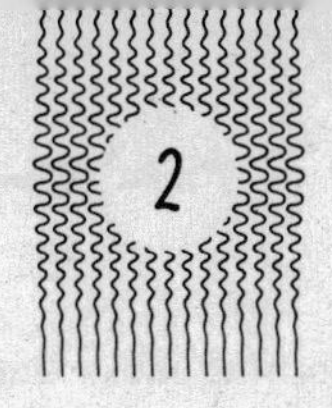

Some Nits, Picked

It all started the day before my birthday. Now that I was married, my in-laws wanted to take me out to dinner to celebrate my turning one year older. We went to a nearby Italian restaurant called Supper. My husband's brother and his wife joined us with their two kids, Alana and Adam. The place was lit in that wonderful New York way where you can barely make out who is sitting next to you. I could mistake a Pilates ball decorated with a beard and curly hair for my husband. You have to use sonar to find the bread basket. It's the best atmosphere for pimples.

After dinner, we walked several blocks to get ice cream at a place called OddFellows where they hand-press their own waffle cones. The smell—sweet and delicate—is exactly how I imagine the scent of Betty Crocker's armpits.

I do not often interact with kids; they scare me because they'll look at you and say things like "Why is your nose crooked?" or they will smile, stare straight into your soul, and then say something creepy like "You're going to be dead."

I constantly judge if I want to have kids depending on the kids I observe around me. When I think of the kid in *Jerry Maguire*, the one who has glasses and that adorable case of asthma, I want to get pregnant immediately. But when I see real human kids who aren't reading off a script in a romantic comedy, I usually want to tie my tubes.

My mom, who wants nothing more than for me to proliferate her genes, knows this about me and gets very concerned if there are kids around us who are misbehaving. She will say, "It's different when they are your own." In the past year, she has grown even more sensitive to crying babies than I am. Once, we were at the grocery store looking at Triscuits and she said, "You know, it's different when they are your own."

"Well, we can buy them," I said. "They're on sale for $2.95."

She cocked her head sideways, and it was only then that I noticed the distant wails from the dairy aisle.

But on this particular evening, things were different. The kids and I ordered the same flavor of ice cream—sprinkles. This made us feel bonded. Alana linked her arm through mine as Adam hooked onto my other arm. They did not look up and say, "You have hairy nostrils." Instead they smiled and giggled as we walked in tandem. They yelled, "Let's walk faster! Faster!" We powered forward, weaving around the crowds, leaving all the other adults behind.

It was one of those rare moments when I thought, *I could do this: I could have kids.*

After ice cream, everyone headed back to our apartment for a final cohesive farewell. When the kids entered, they wanted to sit on my aqua-colored velvet sofa chair. This was not a hand-me-down. This was my first and only real piece of adult furniture. And I went big. Again, in case you missed it, we're talking about an aqua-colored velvet sofa chair.

I finally understood why my mom got so upset when people called her first luxury car—a LeBaron convertible—beige. "It's actually champagne," she'd correct them.

Every piece of furniture I had before this piece—it was so elegant that it certainly qualified as a "piece"—was inherited from the street, and the only other sofa in our house was a little maroon number that I'd guess was about fifteen years old and probably hosting the plague.

The new sofa chair felt all the more precious because we almost lost it before it made it into our apartment. In the spacious furniture store, the chair had looked tiny, like an ottoman for gerbils, but after lugging it up two flights of stairs, the deliverymen found that it wouldn't fit through our doorway.

"Fucking shit," shouted one.

The other one wiped sweat from his brow. "Jesus Christ. This again."

Apparently, it's not uncommon for people's furniture fantasies to be much bigger than their apartments.

"Did you even measure it?" asked the deliveryman who most looked like he wanted to break my face.

"Yes," I said. I left out the "with my mind" part.

I'm usually great at spatial stuff. I can look at a pot full of leftover soup and then select Tupperware to match the amount to perfection. It is one of my greatest gifts.

Johnny, the super in our building, eventually saved the day by taking off our front door, giving us a critical extra two inches. To put our door back on, he then charged us sixty dollars. I got upset with him for price gouging, but then I remembered that time he retrieved a hairball the size of a llama out of my shower drain and that reminded me that he should be given the Medal of Honor and be added to Mount Rushmore.

I mention all this only to explain that on the fateful night that the children arrived, I had been experiencing inappropriately strong protective and possessive impulses toward my chair. Ever since the chair was delivered several days earlier, I'd had trouble sharing it with even my husband. It seemed unwise to have something so fluffy and pristine touched by too many asses. What I once viewed as abominable—the plastic sofa cover—I now thought of as a brave and courageous choice made by grannies the world over. What a beautiful ancient practice!

So when the kids walked in and wanted to plant their butts on my chair, I felt a lot of resistance, but we'd also just returned from such a lovely evening together. I had been so engrossed by our jaunt back to the apartment—they were so fun—that I'd barely even noticed New York City's classic eau de parfum, a bouquet of rotting rat corpse melded with stale urine, which was constantly brewing on our corner.

Also, and most compelling of all, my husband was giving me his famous and highly effective death stare. The only thing that was missing was a red laser beam shooting out from each of his pupils. He could see it in my face, in my posture, that I didn't want our niece and nephew to sit in my new chair, and he did not approve of that inclination.

Reluctantly, I gave the kids permission to sit down.

They sat for a moment—scooted around—but then they quickly became bored. Part of me was relieved that they exited the chair without defiling it while the other part of me was offended that they got over the revelatory seating so fast. I sat down in their stead while they returned to the living room rug, where they dismembered Mr. Potato Head.

We said goodnight and then my in-laws left.

The next day was my actual birthday. It is the one day out of the year that I have profoundly unreasonable expectations for how I should be treated. Logically, I believe it is a gift to be on this planet and we should all spend the day of our birth picking up litter, but something comes over me and I become a complete beast. I feel terrible for my loved ones. By evening, my husband is usually calling me the Birthday Maranster (Mara + monster = Maranster). I even get upset at inanimate objects. Red traffic lights piss me off. Do they not realize that on this day many years ago I came out my mother's womb and therefore, in my presence, they should turn green?

I don't know how I came to feel so entitled. The only thing my parents did for my birthday was let me choose what we were having for dinner. I always picked poached sole over steamed rice with a splash of Knorr instant hollandaise sauce. Besides that, it was business as usual.

I took the day off work, because clearly no one should have to work on her birthday. My friend Maggie took me to a spa called Aire Ancient Baths. The atmosphere in that place is sex dungeon meets Spanish conquest. The bare brick walls, candlelit chandeliers, and stagnant bodies of water made me think of medieval genocides and cholera outbreaks, but in a romantic time-travel way.

We spent two hours shifting from hot tub to sauna until we got hungry for lunch. We went to the locker room to shower and change. I finished before Maggie; she has so much hair that it takes an hour to blow-dry, while mine—though prolific all over my body—is so thin on my head that all I need to dry it is for one ant to exhale onto my head. Instead of getting annoyed at how long she was taking, I checked my phone for birthday messages. I selected the text from my brother-in-law. Even though we had celebrated with them the night before, I suspected he wanted to wish me a happy birthday on the actual date.

"It was nice to see you guys yesterday and spend some time together. FYI both kids had lice in their heads. Didn't know until we got home last night. Sorry about that!"

Uh, what?

I'd been married for a while, so I wasn't sure if things had changed, but I thought protocol was to call, not text, if you'd possibly transmitted a disease.

I read the text again. I tried not to overreact, but this was happening to me on my birthday!

I thought back to how I'd been so reluctant to let the kids sit on my aqua-colored velvet sofa chair. I wasn't an asshole. My sixth sense had intuited that they were contaminated.

I needed immediate mental health support. I got Maggie's attention by yelling above the roaring dryer. She is an inherently positive person. She could see someone lose a leg in a car crash and still probably find something uplifting to say about the disaster. "At least her leg is still intact!"

She'd say it like dismembered extremities have been known to go on to get postgraduate degrees and lead productive lives.

She sat on a bench while I paced back and forth, recounting all the possible points of physical contact I'd had with the kids the night before. I mentioned the dark restaurant, the ice cream shop, and how we'd skipped back to my apartment. At the time, it had seemed like such a sweet and quaint *Sound of Music* moment, but now I wondered if it was all a ruse—the lice made them do it in order to get close to my head. And to think, after that night, I'd almost been convinced that I wanted kids.

"You won't get lice, Mar," Maggie said.

By this time, we were already outside, waiting to cross at a light that apparently didn't realize that it was my birthday.

"How do you know?"

She reminded me that I'd hardly touched the kids. I had to be in closer contact. Lice don't fly. "Did you rub heads?" she asked for the fourth time.

"No."

"Then see," she said, "you're fine."

By the time I met up with Dave for my birthday dinner, I was ready to commiserate with someone, but he wouldn't cooperate.

He's usually the one concerned with external dangers—murderers, burglars, and tiny fanged creatures—while I worry more about internal dangers like cancer and the possibility that my heart may explode. But as I shoved birthday enchiladas into my mouth, there was a serious role reversal at play.

"Don't worry," he said.

"But. There. Were. Lice. In. Our. Apartment," I said for the eighth time.

I imagined the little insects were exactly like criminals, but instead of wanting to hurt or steal from us, they wanted to eat our heads.

"We'll be okay," Dave said.

That night, I felt tingling on my scalp. The sensations seemed real, yet then again maybe they weren't. I nudged Dave awake and asked him to investigate. He rolled his eyes, but humored me nonetheless, because we took vows, dammit.

As Dave positioned me under a lightbulb and parted my hair, I was reminded of lice-check days in elementary school. The first time was in kindergarten. The nurse came in with latex gloves and told us all to sit cross-legged on the carpet. She directed us to lower our chins to our chest and wait patiently until she got around to each and every one of us. I was terrified about what she was going to do until she stood behind me and laid those virtuosic hands onto my head. My eyes rolled back and my arms got gooseflesh as she parted my hair into tiny chunks and looked through each nook and cranny of my scalp. Lice check, I thought, should last forever. Parasite detection felt so good.

From that day forward, I got excited whenever a nurse appeared with latex gloves (an impulse that would eventually fail me).

In the meantime, my friends and I began to play "lice check" during slumber parties. We would take turns being the nurse and the potential parasitic host. I cannot quite imagine what my parents must have thought if they overheard our conversations. "But now it's my turn—you have to check me for lice!" They probably thought I wanted to become a nurse when in actuality I wanted to grow up and get massages.

As I got older, I experienced a new kind of hair caress—the kind attempted by boyfriends. I always wanted to tell them that they weren't doing it quite right. Hair caressing should have direction, a point of view. Their method, I realized, lacked purpose. I wished I wasn't so shy and could have just told them exactly how I wanted it: "Do it to me like you're checking for blood-sucking insects!"

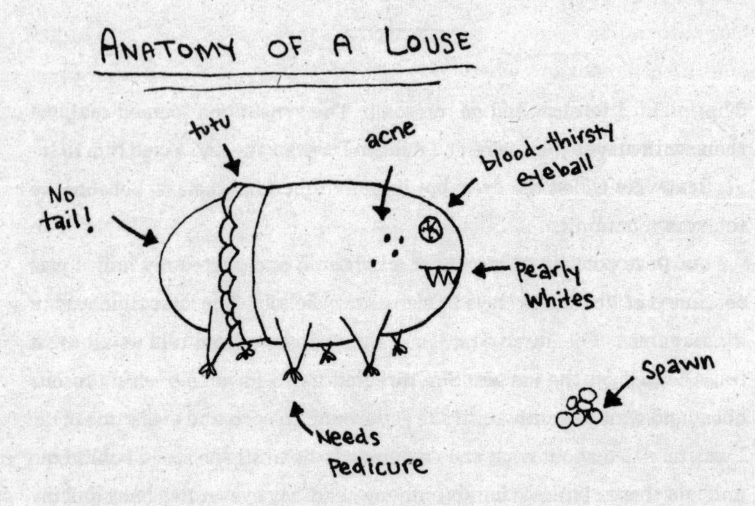

What I'm saying is, I totally understand how childhood can influence the development of sexual fetishes.

"I don't see anything," Dave said, turning off the light.

I spent the next two weeks under orange alert. I didn't necessarily think I was going to get lice, but I wasn't going to be stupid and ignore the fact that lice-infested children had been in my home. I took precautions. The aqua-colored velvet chair was under quarantine as a potential hot zone. I didn't care that science says that it's highly unlikely to contract lice in any other way than direct head-to-head contact.

I started out learning novice stuff—a louse can lay six eggs a day, it can survive underwater for several hours, and infestations occur most often at schools in September (and September it was) because children come back after a long summer break and immediately mingle their head fauna—but I eventually got into mating practices. I dug way too deep. This info wouldn't help me, but I couldn't stop. It was ghastly stuff. If a louse dies while copulating, then the pair can't separate. The one that survives has to carry around the other's dead body, connected via the genitals, for the rest of its life. I didn't know whether to think it was tragic or beautiful—surely dying in each other's orifices was more romantic than in each other's arms, but still.

I kept going down the information black hole. When a louse needs to eat, your head becomes a real-life nightmare. A small tube with teeth on the end protrudes from its mouth and pierces the scalp. As if that weren't obscene enough, the tiny menace then spits on the wound. The spit is what makes some people itch, but it's also magical and keeps the cut from clotting, so the louse can endlessly consume the blood through the two pumps in its head as if it were standing under a never-ending soda fountain. Lice take four to five meals a day, during which they consume the equivalent—if they were our size—of ten gallons of blood, and even with all that liquid, they don't pee. Their urine evaporates through their respiratory system while their excrement, tiny dry pellets, goes through the more traditional anal route.

I later spoke to Kim Søholt Larsen, an entomologist from Denmark with a PhD in fleas and a specialization in lice and ticks, about this behavior. "If they urinated, your hair would stick together and you would immediately figure out that you have lice," said Larsen. "This is how they hide themselves."

They've had a lot of time to hone their terror techniques, because they've been hunting our plasma for millennia. The only good thing I found out about head lice was that they aren't body lice: Body lice carry disease.

In the meantime, I still had Dave check me at the slightest provocation. If I felt anything, I'd turn on the bright overhead lights, flip my head over, and have him gander at my scalp. We were searching for insects that looked like black sesame seeds; irritated red skin near the ears and neck; and tiny white dots—lice eggs or nits—that informational lice blogs described as looking like "dandruff that won't move" near the base of my hair.

I did not moan from pleasure during these encounters. It wasn't that Dave was bad at lice checking, but I found that the practice didn't feel as hedonistic when it wasn't recreational—nothing like a real fire to take all the fun out of a fire drill. Dave would toss a couple of strands here and there and tell me that everything was going to be okay.

"There's a very low possibility that you got it," he said, over and over again. "The kids were here for like two minutes."

During that time, I got so invested in looking for lice that I forgot about my usual terror of tumors. In some ways, it was kind of nice to mix up my concerns. Tumor hunting gets very one-note after a while.

Life continued. If I felt an itch, I made Dave look. Otherwise, I was content just to have a valid reason never to let anyone ever sit in my aqua-colored velvet sofa chair again.

By the time ten days rolled around—which was the amount of time it would have taken any new eggs to hatch—I'd probably had Dave check my hair about fourteen times and we hadn't found one louse. Finally, I felt confident that we had eluded the little bastards and that we were in the clear. It was perfect timing, too, because we had only two days to prepare to take off for our honeymoon.

Dave had had a hard time getting time off work, so we had waited two and a half years after our wedding to take the Japan honeymoon of our dreams. I had spent three months planning the affair. Over eleven days, we would be visiting four bustling cities. We began busying ourselves with packing and plans of what we'd eat.

We arrived giddy and exhausted at Narita airport. Over the next few days, the stresses of the last couple of weeks completely disappeared. I even eased up on the idea of an elective hysterectomy. We became fully engrossed in our new surroundings. We went early in the morning to Tokyo's Tsukiji fish market, and we visited Shinto shrines and Zen rock gardens.

After a few days, we took a train to Hakone, a small mountain town renowned for its hot springs. We stayed in a *ryokan*, which is a traditional Japanese inn. Our room was beautiful and highly flammable. The whole thing was made of wood and bamboo tatami mats. In fact, tatami mats are how they measure the size of rooms in Japan. They don't say a room is twenty by thirty-five feet; they will say something like "It's seven tatamis." (In the United States, we don't have a form of measurement that's nearly as charming. If we tried, it would turn into something awful like "My house is three cement trucks and a granite countertop that caused four people to lose their fingers in a Brazilian quarry.")

A soaking tub was on our balcony. Dense green foliage gave us privacy. We wore robes at all times of the day, because they even brought us our dinner.

Soon we were taking bullet trains, regular trains, and a ferry to find our way to Naoshima, a small island with a phenomenal hotel inside a modern art gallery. In the morning, the seventh of our trip, I went for a walk by myself to admire the magnificent sculptures on the grounds, such as the massive polka-dotted pumpkin by the artist Yayoi Kusama. I continued walking along the shore, ankle-deep in water. I looked out over the serene and vast horizon and felt so much gratitude.

I then noticed that I was scratching the back of my head. *How long had that been happening?*

After taking a shower, I roused Dave and we went to breakfast. "I think I'm having a reaction to the shampoo," I told him over eggs and miso soup. I figured that using all the different shampoos at the different hotels was making my scalp feel irritated.

"Yeah, probably," he said. He also mentioned that it was unusually humid for us so it's possible that I was having a heat rash.

"Yeah, you're right," I said. "The climate really is different here."

He told me that the psoriasis on his scalp was acting up, too.

"That makes sense," I said. He always had more trouble with his psoriasis when we traveled.

Later that afternoon, we made our way via bullet train to Kyoto, where we would spend our final three days. After that, we would have one night in Tokyo before heading back to New York.

When we reached Kyoto, we were exhausted, so we grabbed a quick dinner at a *tonkatsu* place before going straight to bed. When we woke the next morning, we both had a cough and runny nose. Nonetheless, we explored Ninomaru Palace and Kinkaku-ji temple and attended a kimono fashion show.

By that point, Dave and I were both having coughing fits, but my back also itched painfully. It was an odd symptom for a cold. I got onto the internet and typed in, "Why does my back itch . . ." And then the search engine autocompleted ". . . when I cough?"

That was comforting. I obviously wasn't the only one. The explanation I found said that when we cough, the nerve fibers in the diaphragm can become irritated by overstimulation. Because there aren't a lot of nerves in our organs, the brain gets confused—there is a crossed signal of sorts—and makes you feel like your back is itchy when it's actually not.

The takeaway: My itchiness was clearly an illusion.

That evening, we were so bad off that we went to a pharmacy. The two-story shop was floor-to-ceiling packed with fluorescent boxes. No one spoke English and none of the medicines had English translations. They weren't even in Roman letters. If they were in Spanish or Italian, I could have at least tried to pronounce the words and then pretended that I knew what they meant. But with Japanese symbols, I was so hopeless that I might as well have been trying to read a pile of pick-up sticks.

After a half hour, we gave up and bought two mystery boxes of drugs. For all we knew, they could have been to treat a dog's case of heartworm and to give me an erection for twelve months. We brought the medicine back to our hotel, knocked back a couple of gel caps, and sat watching the news in Japanese. An hour later, I was still scratching. During a commercial that depicted a woman having an intense flirtation with what looked like a fried chicken cutlet, I looked over at Dave.

I stared at him until he said, “What?”

“Do you think it’s possible that the itching is from lice?” I said.

“I doubt it,” he said.

It did seem unlikely. It had been almost a whole month since we’d seen his niece and nephew. If I’d had lice, wouldn’t they have made themselves known weeks earlier?

I asked if he’d check just to be sure. The light wasn’t great in the room, so I sidled up to the bedside table lamp. I turned my head upside down as he looked through my hair.

“I don’t see anything,” he said.

I was pleased with that answer, so I turned over and went to sleep.

The next day we had a cooking class and then switched from our hotel to a *ryokan* in town. We wanted to get a little more of that traditional Japanese feeling. Despite being ill, we went for the multicourse *kaiseki* dinner. Between the pickled vegetable course and the fish stew course, a small brown bug fell onto Dave’s arm. He immediately flicked it off.

“Where did that come from?” I asked.

“I don’t know,” he said.

It almost looked like it had dropped from his head. There were a lot of quirky things in Japan. A raccoon dog, called a *tanuki* and known for its colossal scrotum, is supposed to bring good fortune. Replicas of the big-balled animal greet you at the front door of many restaurants. In a country that adores a rodent with gigantic testicles, why wouldn’t a bug appear out of nowhere?

After our final day in Kyoto, we headed back to Tokyo. We had dinner at a sushi joint in the Ginza district. I itched so badly that I couldn’t keep my hands out of my hair for more than one slice of fish. That night, I couldn’t sleep, so I went to the swimming pool as soon as it opened. For the first few moments, the water put out the bonfire of pain on my back and head. When I got back to the room, I had to ask Dave to check for lice again.

He still didn't see anything, which was actually a huge relief, because if I had lice, it would be somewhat of a cataclysmic event for Japan. I had been up and down the country using the bullet train. I used blankets, pillows, towels, taxies, ferries, and small Jacuzzis. I laid on tatami mats and rubbed up against restaurant booths. I had leaned against walls, tried on *yukatas* (thin cotton kimonos), and wrapped cute scarves from expensive shops around my head.

"It's probably an allergy," Dave said, which sounded entirely plausible even though I'd never once experienced an allergy.

"Yeah, we'll figure it out when we get back," I said.

By this point, we were ready to get home. The same man who'd picked us up at the airport eleven days earlier drove us back. I noticed how he'd decorated his car headrests with intricately woven lace doilies. So many people in Japan went the extra mile to make everyday objects more comfortable and aesthetically pleasing. I was impressed. I laid my head back onto those beautiful covers as I watched the city go past.

At the airport, we had a little extra time, so I went into a corner shop. I browsed books and then I began trying on neck pillows. My mom always told me not to try on stuff like that in stores because you never know the hygiene of other people who have tried them on before you, but I've never been concerned. The pillows were so soft and came in so many colors.

As I tried them on, I became a little obsessive—it happens periodically—and suddenly felt like as long as I tried on every different color, then somehow that would mean that the plane wouldn't crash.

Dave was getting antsy, but I managed to finish my mission before he dragged me off to our gate.

When we got home twenty hours later, we went straight to bed. I woke up on a glorious Sunday morning, and the first thing I did was jump into my aqua-colored velvet sofa chair. I could once again enjoy that plush swiveling piece of gluteal glory, because it was finally out of quarantine.

After fully indulging, I started to unpack our bags, piling our dirty clothes onto the other sofa. While I was doing that, Dave woke up and suggested that we go to the farmers' market. We'd been eating gluttonous meals for the past eleven days and he thought we should get some fresh veggies.

I left our clothes strewn in the middle of the room as we went out into a chilly but sunny New York morning. We walked together in the East Village along Avenue A, up toward St. Mark's Place. We were talking about what we would make—some kind of soup? No. A roasted chicken? Maybe. Something with black beans? That sounded good.

I remember happy dogs walking by with their owners. The clank of boots on the sidewalk cellar grates. Pulling my sunglasses down over my eyes. The burn at the back of my head. The stinging sensation that occurred each time I touched my scalp.

I stopped in the middle of the sidewalk. The vendors—their piles of gourds and apples—were in sight.

"You have to check my head one more time," I said.

"Right now?" Dave said.

I didn't answer him. I didn't even care about standard pedestrian practices. I stayed put in the middle of the sidewalk, like an obstinate boulder dividing a rushing river, as people walked around me. I dropped my chin to my chest and waited until Dave appeased me.

When I was in Japan, I could easily dismiss the sensations as if they were some kind of awkward travel bug—the customary stomach upset we expect when traveling to a new place—but now that I was back home, I could finally recognize that the shit I was feeling was not even close to normal.

Something had to be wrong.

By this point, playing lice check had lost all its former cachet. Dave was exasperated—he'd probably checked my head at least thirty times—but he did his duty and took his designated position behind me. My hair was in a bun, so I expected him to start rummaging around in there. Instead there was silence and the heat of direct sun.

"Do you see anything?" I said.

"Um," he said.

"What?" I said.

There was another long pause.

"What?" I said.

He came back around to face me. The corners of his mouth were drawn down. "It must be because there's better light here," he said.

Crabs Pubic lice, or *Pthirus pubis*, are the couch potatoes of the lice kingdom. They are characterized by their sluggish and sedentary lifestyle. I can't blame them; I'd be that way, too, if my house was a porn set. Each louse is a millimeter, which means it would take twenty-five of them, back to front, to add up to an inch. They have a roundish gray body with six legs. The two in back are capped off with crustacean-looking claws, which is how they got their nickname: crabs.

They are not found in the crotch because they are fools for genitals, but because pubic hair is their method of transportation. Like a train needs tracks to move, crabs need pubes. That's why they can also be found in other coarse hair like eyelashes, eyebrows, armpit hair, and beards. We originally caught pubic lice from gorillas three or four million years ago. That's why pubic lice like pubes. Pubes are the closest thing we have to thick and tough gorilla hair. The fine hair found on our scalps does not give them enough purchase to move around.

Crabs don't do much besides suck our blood and lay eggs—about three a day—for the two to three weeks of their short lives. Like head lice, they can't jump or fly but can only scuttle from hair to hair. That is why sex—pube to pube—is their best opportunity to colonize a new home. They can also,

"What do you mean?"

He told me that there were so many black sesame seeds moving around that he couldn't even count. He said it looked like a horror film where bagel toppings came to life.

My first reaction was to laugh. Gosh, isn't that funny. I have a lice infestation. I went through an entire country spreading a parasite during my honeymoon. LOL!

Uneasily, Dave joined in on the laughter, too.

Then we pretended that whole episode didn't just happen. We continued walking toward the farmers' market as if we were different humans—ones who didn't currently have minuscule animals eating away at their

though extremely rarely, be caught through infested bedding. A myth looms large that crabs can be transmitted via a toilet seat, but if that's how your boyfriend is telling you he got his, then it might be time to find a new boyfriend or to finally have that talk about opening up the relationship.

One textbook, *Medical Entomology for Students*, explains quite insightfully that having lice makes one "feel lousy." Crabs can cause itching and irritation, but they are also easily exterminated: Wax off your bush or use insecticides.

Though crabs—blood-sucking wingless genital goblins—sound apocalyptical, we actually have them on the defensive. They are becoming endangered because of habitat destruction. In one study, "Did the 'Brazilian' Kill the Pubic Louse?" researchers found that the dwindling number of crab infections coincided with the wax-it-all-off trend, which began around 2000. It's hard to get good data—people often don't report embarrassing parasites that have staked out their perianal region—but a 2009 study from East Carolina University reported that less than 2 percent of the population harbors *papillon d'amour* (which is the sexy French name for crabs). "Their forests are disappearing," Danish lice expert Kim Søholt Larsen told me. "They are endangered because they don't have anywhere to live."

flesh. It was the most acute case of denial I'd experienced since I was twenty and still suspected that I might grow another ten inches.

"So we're going to get broccoli and what else?" I said.

"I don't know," he said. "We'll have to see what else looks good."

We were half a block away from the vegetable stalls when we both paused and looked at each other.

"Wait, we can't go to the farmers' market right now," Dave said.

I furrowed my brows as the realization finally dawned on me, too. "Holy shit," I said, "I have lice!"

An hour later, I was sitting on a chair in our apartment hallway. Dave stood behind me, brushing through each segment of hair with a fine-tooth comb. We had bought just about every lice-murdering product at Duane Reade, and upon getting home, I had doused my hair with the toxic shampoo. There were nontoxic methods, but I wanted poison! I wanted complete decimation! The fumes—strong and searing—were making my eyes sting, and I relished the implications of this particular burn.

Dave sounded bilious as he explained the scene he was confronted with: "It looks like a city was napalmed and the civilians are trying to escape." Many lice ran down my back. I couldn't count them all, but I'd guess there were at least a metric shit-ton. On a piece of paper towel, Dave showed me an abnormally large one. "Look familiar?" he said.

It looked exactly like the bug that had fallen on his arm in the Kyoto *ryokan*.

(To this day, that bug is still inexplicable. I looked it up and there is no such thing as a queen louse. I try not to wonder about that too much. Mostly, the lice were as billed: dark brown and the size of sesame seeds.)

While I sat there, I thought back to all the neck pillows I'd tried on at the Narita airport. I wondered if lice inject you with psychotropic substances that make you think it would be a great idea to rub your head all over everything. (I'm sorry, people of Japan!)

Dave, oddly enough, had only four lice in his hair. When we did some research, we found out that they were repelled by the acidic shampoo he uses for his psoriasis. It was nice for him to realize that there was at least one positive to having a skin disorder.

Even though I didn't tell him at the time—it was my duty to make him feel guilty for being a subpar lice-checker—committing genocide on my lice population was one of the most romantic things that he'd ever done for me.

I didn't speak about my parasite to many people, because having lice is stigmatizing and they scare people, as they damn well should: Those suckers hurt and they are immensely contagious from head-to-head contact. Those evil little bastards exploit our love of hugs. That's how they've survived for like a billion years. Nits have been found on Egyptian mummies. Vikings even carried delicately crafted lice combs in their belts alongside their most essential item: their sword. They—muscular masculine marauders from Scandinavia—were so freaked out by the little bugs that they got buried with their combs in case they needed to battle lice in the afterlife.

One of the few people I told was my dad. "Oh, yeah," he said. "You know the story about how me and your mom got lice, right?"

He was referring to the time they both got crabs when they were twenty. Even though I've heard the story several times, I still don't know it, because I've worked hard after each telling to block it out.

"Dad, I didn't get genital lice!" I said.

He told me that was too bad because it meant that my lice story was a helluvalot less interesting than his.

When I got off the phone, I spent the next day wondering how lice knew which patches of hair they belonged to—did I have to worry that my head lice could suddenly, due to positioning, become pube lice? Luckily, I found the answer to that was no. Head lice can move around only on thinner head hair, while pubic lice evolved to navigate coarser hair.

Except for one particularly bad day when I contemplated lighting my head on fire, I brightened up over the next few weeks. I also stayed incredibly vigilant. If you leave one louse or nit behind, you can easily reinfect yourself. I knew it was overkill—many entomologists say you cannot catch lice from anywhere except head-to-head contact—but because I'd spread all my infested baggage all over our apartment as soon as I'd gotten home from Japan, most of our place was under quarantine. That, of course, included the increasingly superfluous aqua-colored velvet sofa chair. I did find humor in the fact that something widely considered a childhood affliction was preventing me from using the piece of furniture that symbolized my burgeoning adulthood. It felt like someone, somewhere, was trying to sabotage my maturity.

After two weeks without any evidence of lice or nits, one is considered in the clear. Until that time, I kept up a daily routine. Every morning, I'd wash my sheets, shampoo my hair, comb it out with a tiny-tined comb, and then investigate any detritus with a magnifying glass.

During this process, I came to realize that if it weren't for me, then all those tiny beings wouldn't have had life. I gave them life. They gestated near my follicles, hatched from my strands, and "breast fed" from my scalp. They could not survive without the heat from my head. You give and you give. They take and they take. Throughout it all, you worry nonstop. Is this what it feels like to be a mom?

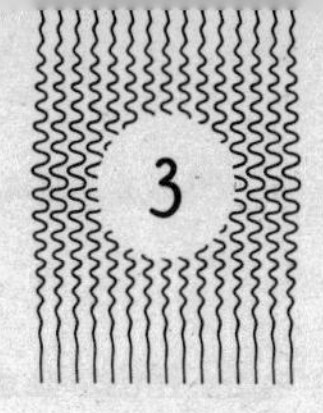

Face It

A friend once told me that I look exactly like Matza Ball Breaker, a girl on the Chicago roller derby team. She called our resemblance "uncanny." So I searched for Matza Ball Breaker on the internet. When I saw her, I was mystified. We both have hair on our heads and a chin below our mouths. We could also both claim a set of eyes. Most likely, she, like me, had a vagina as well. Other than that, I was left deeply confounded. My supposed doppelgänger looked nothing like me—or at least the concept of me that exists in my head.

Even though I've seen my image—photos and reflections—for thirty-four years, I'm confused as to which—if any—portrays reality. How I appear to myself is not at all consistent; my image is like a moving piece of newsprint that I can never fully read.

The me that I see in the mirror is often, though not always, more attractive than the me I see in photographs. When I see photos, it feels as though I must have been kidnapped as the shutter tripped and had Yoda placed in my stead.

There is nothing worse (except for murder, of course, and finding a long wiry hair in your entrée) than hearing someone say, "That's a great photo of you," only to get a glimpse of it and see staring back at you a mustachioed gnome with water-balloon cheeks and a grimace that could stunt-double for an elephant's anus. If that is a "great" photo, then what am I when I'm captured at my everyday?

I've tried to get a handle on this discrepancy by pointing at various disagreeable photos of me and asking my friends, "Is that really what I look like?"

Then it's often discouraging when they say, "Yes."

In order to survive, I have to tell myself that everyone—all the people in the world—must be way overdue for cataract surgery.

No matter how sure I am, my perceptions are inevitably challenged. Recently, I snapped a selfie that I liked—*there I am*, I thought after taking my twenty-fourth shot—so I asked Dave for validation.

"How about this?" I said full of hope. "Is that what I look like?"

"Yes," he said.

I was elated until he quickly added, "Except for your face is much rounder and your cheeks are bigger."

Thus, the various manifestations of my appearance continue to confound me.

I was always uncomfortable with the author photo on my first book, but not for the usual reasons. This photo actually promised a little too much—unlike most, it didn't make me look entirely like an Ewok—but friends and family reassured me that it was a fair depiction.

I spent many months going to events with a fear that I'd sense a palpable disappointment upon the audience's realization that the real me didn't live up to the poster outside. Everything was okay—if it was happening, people kept their snickering to themselves. I felt encouraged—perhaps I actually *was* attractive—until a loathsome evening in the middle of June at a small event space in Midtown Manhattan.

Before the reading, a woman lingered in the back by the table of books. She had my book in her hand and was rifling through the pages. She nonchalantly asked me who I'd come to see.

"I'm actually reading tonight," I told her.

"Which book?" she asked.

I told her it was the one in her hands.

She turned the book over and appraised the photo. "Oh, that's you?" she asked. I sensed a bit of incredulity.

"Yes," I said.

She laughed and gave me a knowing glance. "I have some glamour shots, too," she said.

After all the evidence—the misleading doppelgängers, the fickle photos, and the many unreliable reflections that chase me around the city in storefront windows—all I can say about my appearance with any certainty is that I have brown hair, a mouth, and a couple of ears. I've been wondering about it for years, so I finally wanted to know, why is it so difficult to get a real read on our own appearance? Is there a true version of the self, and if so, can we ever see it?

At first, I suspected that the inconsistency I experienced with my looks was solely an issue with the medium I used to view myself. There was something mysterious that happened—I became uglified—when my image hopped from a reflection to a photograph. Cameras, those bastard devices, had always misunderstood me.

To fill me in on what might be happening, I spoke with Pamela Rutledge, the director of the Media Psychology Research Center. She said what many of us might already know: The mirror is a small white lie. It flips our image. Unless our faces are perfectly symmetrical—which happens only in the rarest of supermodel cases—we will likely feel uneasy when we see a photograph of ourselves. The nose that usually leans to the right in a photo leans to the left.

"It can look slightly off and therefore look funny to us," Rutledge said. She explained that many of us prefer our mirror self simply because we see it more often. "We like what's familiar," she said.

"We like what's familiar" sounded like an off-the-cuff generality, but it's actually science. We tend to develop a preference for things—sounds, words, and paintings—for no other reason than that we are accustomed to them. This concept, called the Mere Exposure Effect, was proved in the 1960s by a Stanford University psychologist named Robert Zajonc. (Finally, there's an answer to the shoulder-pad craze of the 1980s. Just by being repeatedly exposed to something—even if it's heinous—you can come to think of it as a good-looking fashion statement.)

Another issue with the mirror is that we all, unconsciously, shift into flattering positions—hide the double chin, suck in the stomach, pop the hip—but it takes only one candid photo to haunt all that hard work and make you second-guess everything. I thought my arms were svelte little hockey sticks until a camera came along at an angle I was unaccustomed to and captured them in a way that gave me a month of night sweats.

Photographs, like mirrors, also don't tell the whole truth. Depending on lighting, focus, and lens size, they can distort us in various subtle ways.

"So which is more truthful?" I asked Rutledge of the two mediums.

"'Truth' is such a subjective word," she said. "A mirror is going to feel more comfortable to you, but a picture is how other people see you."

"Gross!" I said.

I thought about this for a while—what Rutledge had said—and decided that it was unacceptable to me. I did not look like my photos. It just wasn't possible. I needed a second opinion.

So next, I got in touch with Robert Langan, a psychologist who dabbles heavily in theory of the self. I went to his office on the Upper East Side and sat across from him on a sofa, the one where so many of his patients have probably cried about extramarital affairs and admitted to being bronies. I still wanted to figure out what I really looked like and if I could ever see that real me. I'd warned Langan that I was coming for that reason, but the situation got out of control immediately. We passed rational query right by and went deep into existential quandary.

"Certainly my nose is part of me, but am I inside of it or on the other side looking at it?" he said of the face we see in the mirror. "Me is inside, but also outside. Outside inside. Inside outside."

I needed to rein this guy in. I asked him how I could get a grip and see the true version of myself. "I just want to see what is," I said.

As he spoke, he tented his left fingers onto his forehead as if his hand were extracting thoughts from his mind. "If you try to see what is," he said, "then you begin to see that you can't see what is, because look at the two hundred and fifty-two expressions you can make in the mirror within a minute."

He raised his left eyebrow. "So which one is you?" he asked provocatively.

He made a good point, but I had a good answer. I told him that, obviously, out of the two hundred and fifty-two faces, I would identify as the most attractive one.

"If the ugly face is false," he said, "then so is the pretty one."

It was starting to seem like all these theorists were hell-bent on traumatizing me.

First, Rutledge told me I look like my photos and now Langan was telling me I look like the me—face twisted in anguish and grief—I saw when I spotted my first chin hair.

"Dammit," I said.

Why couldn't these people be more supportive?

"You're running at sixteen frames a second," Langan said. "If you take out one frame at a time, each image will have a momentary truth to it, but it's really the flow of it all that makes the difference."

Ultimately, he was saying that there is no way to see the self, because we each have many selves, all of which are equally true. I could never know what I look like, because there is not just one me.

Clearly, I needed a third opinion.

This time, I approached the question from a different perspective: I went into the brain.

I found this guy, Julian Keenan, a neuroscientist, who studies consciousness and is also a coauthor of *The Face in the Mirror: How We Know Who We Are*. He led me to believe that seeing ourselves clearly has a lot less to do with our eyes than it does with our minds.

Once I got ahold of him, I got straight to the point. "Is there a way to see ourselves accurately?"

"No!" Keenan said. He said it so emphatically that I pulled away from the phone. "I mean, you'd have to meditate in a cave for twenty years to lose all the old messages about who you are."

He also suggested that a traumatic brain injury might do the job (no thanks!) or a large dose of hallucinogens (maybe!).

Keenan went on to explain a psychological phenomenon called top-down processing. A top-down process is when something already exists in our brains—a belief, an attitude, or an expectation—that affects our perceptions of the outside world. People judge a hill as steeper when they are wearing a heavy backpack. The more depressed someone is, the darker they will rate an image. How we see ourselves—our own reflection—is also colored in this way.

"So that image that you see is going to have thirty-four years' worth of garbage thrown into the backseat," he said, "and each time you see yourself, you're hitting the breaks and the garbage is flying into the front seat."

"Like all the things your cheerleading coach told you?" I asked.

In high school, I was in coed cheerleading, and my coach once told me I'd gained too much weight, that if I gained one more pound I couldn't do stunts or else I'd break a boy's neck. I went through several years of looking into the mirror and seeing only fat fat fat.

"Yeah," he said, "and you've been asking questions about your own face forever. Do I look like my mom? Do I look pretty? Do I look better if I trim my eyebrows? Whatever it is, we've been focused on this thing for a long time."

It seemed so odd: How could I look into a reflective surface—something that seemed so tangible and concrete—yet be incapable of seeing my own image? I wanted to be sure I got this right. "So we're not who we see when we look in the mirror?" I asked.

"No, probably not at all," Keenan said. "Maybe when you were two and saw yourself, yeah, but after that, no. You're just bringing too much to the table."

As it turns out, Keenan was not bullshitting. There is actually a small body of research proving that we suck at seeing ourselves accurately. When I read the studies, I felt vindicated yet unsettled. Weird shit was definitely happening.

Nicholas Epley, a professor of behavioral science at the University of Chicago, suspected that we are terrible judges of our facial reality, but he didn't know to what extent. To test his hunch, he snapped the headshots of twenty-seven people and then altered their photos in 10 percent increments to look both more attractive and less attractive. To do this, he morphed the photos with a composite of hot people and then a composite of people suffering from craniofacial syndrome (misshapen faces). He then had the twenty-seven people come back into his lab and select which photo, out of the eleven, was the real one.

Most selected a photo that was more attractive than that of their actual selves. Yes, these people have known their own faces for their entire lives, yet they were incapable of picking themselves out of a lineup.

I called Epley to get a better grasp on this idea. "So we think we are hotter than we actually are?" I asked.

"Yeah," Epley said, "on average that's true."

That would explain why I get so startled and disappointed whenever I see my actual likeness. I was a 6 in the mirror, but an 8 in my head.

I wondered about the people who are already 10s. They can't get any more attractive. When Gigi Hadid envisions herself, does she see a GIF of a unicorn shooting glitter out its orifices?

Epley explained that this phenomenon happens because of something called positivity bias. "We want to think the best of ourselves," he said.

Studies show that people think they are funnier than they are. They misjudge their ability to be a leader. They also tend to believe flattering information but deny anything derogatory. In this study, Epley found that the higher self-esteem the person had, the more likely that person would select an enhanced photo as her own. It was the top-down process that Keenan had mentioned. Our self-esteem—the idea of ourselves that exists in our head—colors how we see ourselves.

"That means people who have really bad self-esteem might think that they are less attractive?"

"Yes, and that's why psychotherapists are in business," Epley said.

"Do you know why our brains do this to us?" I asked. To me, it seemed like a clear-as-day survival mechanism—up there with blinking and breathing. If we want to have a good time, we've got to go out on the town believing that we are mate-worthy. We have to strut our stuff, even if we don't have stuff.

But Epley wouldn't have any of it. "I think it's probably just a happy accident," he said. "People are very quick to find meaning or functionality or purpose or intention in something when you don't really need it."

In a different study, David White, a face-perception researcher at the University of New South Wales, also found that people are terrible at knowing what they look like. He pulled online images of people and then asked those people to select the image that looked most like them. It turned out that the images they chose were quite shoddily matched to their current appearance. In fact, images that had been selected by absolute strangers bore more resemblance to their current appearance than the ones they themselves had chosen.

"It's quite surprising," White said, "we look at our own face a lot—more than any other face—yet we seem to be quite poor at picking it out."

That explains why Bethany in accounting uses a profile photo on her Facebook account that looks like it could be her third cousin twice removed. If you want an accurate profile photo on social media, White suggested, ask a rando from the street to pick the shot.

To understand why we are so challenged and inept at this seemingly simple task, he said, more research needs to be done, but he had some theories. Along with positivity bias, he believes we look at our own face differently from how we look at other faces. "We look at other people's faces in a holistic context—we see their whole face and all the expressions that go along with it," he said. "But when we look at our own face, we're not really interested in those social signals, but more for grooming purposes."

In other words, we can't see our face for the moles, pimples, pores, and smudged eyeliner.

Another possibility is that we have memories of our face, decades of different representations, which White believes muddle our ability to view ourselves clearly. "If you've been writing something for days, you become blind to the grammatical errors and typos," he said, giving an analogy. "Your familiarity with the text impedes you from seeing it properly."

Unfortunately, this isn't like an essay—you can't put your face in a drawer for a few weeks and come back to it fresh-eyed.

Was there a lesson here? When I asked Epley if there was a takeaway, I thought he was going to say something like "Be the girl boss you feel like you are in your head," or maybe "Don't worry what the mirror says, I just Google-image-searched you and you're plenty slamming."

But for him, this study was a small piece to a much larger puzzle; it wasn't just evidence that we misperceive our appearance, but of how our human minds constantly screw up. "We are not as accurate as we think we are," he said. "Even with something like knowing what you look like—for God's sake you ought to be able to identify yourself in a mirror—but even that judgment is distorted."

He said that humans need to have more humility. "We are wrong more often than you might guess."

"That's the message?" I said. "Really?" I'd expected it to be a tad bit more inspiring.

"Yes," Epley said. "Take a more humble approach to interacting with other people or even thinking about yourself—recognize you're often wrong."

"I didn't see that coming," I said.

"Exactly," he said.

Three days after speaking with Epley, I found myself on a two-hour bus ride up to Rosendale, New York (a town known for its cement), to do what everyone said wasn't possible: get a glimpse of my true self.

John Walter, the man I was going to meet, had invented a device that, he said, would enable me to see myself exactly as I am. Considering that I'd just learned that our brains are terrible at that kind of maneuvering, this was an intriguing turn.

As the bus rolled down the highway, I had my thoughtful face on, which regrettably at times has been misconstrued as constipated, while I wondered what would be revealed.

Was I going to see Matza Ball Breaker staring back at me?

When we spoke briefly before the visit, John had said, "There is a real problem mirrors are causing everyone, but they just don't realize it yet."

HOW WE SEE OURSELVES (on a good day)

How (It feels like) OTHERS SEE US

"How do you mean?" I asked.

"You're trying to figure yourself out," he said of the reflection we investigate in the mirror each morning, "but through the wrong person."

The bus dropped me at a desolate parking lot. John was waiting there in his green Ford pickup truck. The fifty-seven-year-old's bright-orange T-shirt contrasted with his reserved demeanor. He was thin with a manicured goatee, mussed hair, and fingers callused and stained with grease. He seemed capable yet sensitive—my guess was that he'd be the type who, during a zombie apocalypse, would forsake violence and instead build a ten-foot wall and start a community garden.

"If you're open-minded, you're going to see it," he told me as we drove toward his workshop. "You'll see what I'm talking about." He was talking about his invention, the True Mirror, and those who through it experience self-revelation. "If you're closed-minded and susceptible to societal bullshit," he said, "if you're the conforming type, you're not going to like it. No way."

Trinkets hung around his rearview mirror—a moonstone encased in silver and two nails bent into the shape of a heart—gifts from those at Burning Man who'd managed to "see it."

As the trinkets swayed back and forth, I convinced myself that not all would be lost if this visit came to naught, because at least if the *Jeopardy!* answer "This site supplied the cement for the Statue of Liberty's pedestal" ever came up, I'd now know that the question is, "What is Rosendale?"

After twenty minutes, we pulled up to John's workshop, a greenhouse-shaped structure made out of steel, which was situated under the bows of giant trees and a cricket's hop from a small stagnant pond.

He is a computer programmer three days of the week and works on this—his passion project—in all his spare time. He gave me a tour of the premises; it was dark inside and filled with a jumble of supplies: papers, bolts, screwdrivers, saws, cardboard boxes, file cabinets, and small rectangular mirror cutouts stacked side by side.

John grew up in New York City and never liked how he looked. He was unpopular and self-conscious, but when he was twenty-two and high—so high—he went into a bathroom. He stared at his reflection and grew upset. He thought he looked fake and ingratiating. Then he spotted himself, by accident, where a mirror met the medicine cabinet mirror: It was his true reflection. He smiled—smiled so goddamn big. It changed how he saw everything, but mostly how he saw himself. He has spent the last thirty years developing a device that could re-create that moment for others.

The True Mirror, his invention, is made up of two mirrors set diagonally into a box so that they meet at a right angle in the middle. This angle creates a true reflection, meaning you see yourself exactly as you would if you approached yourself on the street. Your left eye stares straight into your right eye. If you lift up your hand, the hand on the opposite side of your reflection lifts. It is a novel way to see oneself, but John thinks there's something much deeper to the experience.

"Check this out," he said, waving me over. As soon as he began talking about the True Mirror, the tightness in his face I'd seen earlier subsided into a full-blown smile—like the pride of a dumpster diver when she finds an untouched burrito.

On top of his workbench, John had unrolled a large canvas poster of Bruce Willis. His face, showing a pleasant smirk, had been deconstructed into parts—an eye here, half of a mouth there.

"Do you see the right side?" he asked, covering the left side of Willis's face. "You see the smile there?"

I saw what he was talking about; Willis looked moderately delighted. "Then you flip it," he said, showing me the flipped version, which was printed below and what Willis himself would see if he were looking at his own reflection. "That same smile now looks like he's being sarcastic with you. It's snarky."

John was demonstrating how when we see ourselves in the "backwards mirror"—that's what he calls a traditional mirror—we might be interpreting our own facial expressions inaccurately. There is not yet any scientific evidence to back up his claims—"I've tried to get scientists interested in studying this, but they won't come around"—but John swears by his own experiential research. And for what it's worth, it was compelling.

"When you flip those little micro-expressions so that they are on the wrong side and the eyes, whatever is in them, are on the wrong side," he explained, as he rolled Bruce Willis back up, "your interpretation is going to differ from what's really happening."

He slid the poster onto a dusty shelf. "Do you see?" he said. "We've been fed faulty information about ourselves for our whole lives!" He attempted to press the import of this upon me even more. "Your face stops working in the backwards mirror," he said. "The feedback loop doesn't work. You stop what you're doing and just stare!"

I think he was referring to the dead look in our eyes—the hunter's determined glare—that occurs when we are about to stalk and unearth a blackhead. I knew that look well.

He put his hand on his hip and took stock of the room. "A lot of the issues people have, especially women, is because all they see is a face; they don't see the bright happy self that's going through the world that other people see."

In the True Mirror, he believes, not only do we see the components that make up our appearance, but more important, we see what makes those parts come alive. "It's that sparkle that comes through," he said.

John was done with his spiel, but I wasn't ready to see myself yet. I felt concerned. What if I'd been waiting all this time to see the real me and it turns out, like the research, that the fake me is better?

Once you see the real you, can you ever reclaim your ignorance?

Besides, I had reason to be worried. One of John's biggest challenges in growing his business—and the reason he still has a day job—is that most people loathe his mirror.

"Fifty percent of people absolutely can't stand it," he admitted. "All they see is where their faces are asymmetrical and they stop there." He said it's gotten so bad that he's had people panic and yell, "Why are you doing this to me?"

"I can go blue in the face trying to get them to see it," he said, "but if they are looking at themselves with a freaked-out face, it's only going to get worse."

Merely 10 to 20 percent of people who gaze into the True Mirror like what they see, while the remaining 30 percent are apathetic. That creates an extreme uphill battle for John's one-man company. He toils in his workshop twenty hours a week and after twenty years of the True Mirror being on the market, he makes only about twenty-two sales each month. He'd told me earlier that even his parents are scornful and puzzled by his mirror obsession.

"What keeps you going?" I asked, when we went outside to get a little air. "You're not making money, and most people despise the product."

"The biggest thing is the idea of making a difference," John said. "Most people don't want their lives to be meaningless—they want to do something that matters, and this matters in spades." He said that the mirror could change lives. He recounted, "After seeing it, I had a woman say to me, 'Oh, that's why people like me.'"

"Can you believe that?" he continued. "She's going around, this gorgeous and generous human, and she doesn't understand why people like her." He told me that another woman, upon seeing her true reflection, said, "Oh my God, I *am* beautiful!"

I like to think I have an evolved bullshit alarm as well as a highly acute cornball siren, and while I knew a lot of this was the stuff of a cloying after-school special, I was smitten by the idea: a mirror that lets you see yourself how your friends see you—to see that they want to hang out with you not because of your perfectly aquiline nose, but because when you're happy, you infect them with your joy.

I had to catch a bus back to the city, so after forty minutes of talking under a tree, John and I walked back to his workshop so that I could see my true self.

From the back of his studio, he grabbed a yellow-framed mirror and placed it on top of a filing cabinet near the only window.

"How do you prep me?" I asked, before taking a peek. This felt like a pivotal moment; I wanted to be sure I was ready.

"There's nothing you're supposed to do—just enjoy it."

John must have been too excited because as soon as I stepped in front of the mirror, he couldn't help narrating my experience. "I can tell that's not a real smile," he said. "You have to be yourself to see yourself."

I couldn't smile yet; I had to get used to the fact that the real me wasn't Gisele Bündchen. I didn't want to mention it earlier, but I'd reserved a small sliver of hope that I was actually a supermodel. So I wasn't Gisele. But I wasn't Matza Ball Breaker, either. I was a person with an exceptionally crooked and miniature beady right eye.

Why didn't anyone tell me? That little obscenity might qualify me for a government check.

John stood behind me, witnessing my rapid nosedive. "But do you see how vibrant your eyes are?" he said, trying to pull up on the wheel. "This is how you talk to the world. Your face works *this* way."

I was in hell another couple of minutes before things shifted. I'm not sure whether I saw it at that point or if I wanted to feel special by being part of the 10 to 20 percent who got it, but I turned that ship around—ogre eye be damned—and began beaming.

"There, now that's a real smile," he said. He told me I was "sparkling" and that I was in a feedback loop with myself, just as I would be with someone I was having a conversation with over dinner.

"You get to see yourself in flow," he said. "That's your energy that you've been missing out on."

I was reminded of something the psychologist Robert Langan had mentioned back when I was in his office. "You're moving all the time," he had said. "Therefore there isn't one truth—you can never look in the mirror and say, 'Now, I'm finally myself.'"

John was trying to show me that—that I was a dynamic and continuously changing person.

"If all you're trying to be is beautiful," Langan had said, "then you're killing yourself. You're getting stuck; it limits you."

I continued to peer into the True Mirror.

"I still think I look weird, but the smiling feels good," I conceded.

"And see how your smile keeps growing," John said.

And when he said that, it happened: My smile grew.

I went full-out. I started making faces at myself. I caught sight of goofy me, the one who makes my husband roll his eyes (out of love and devotion, of course) and the juvenile me who searches for a face to share a giggle with when someone inadvertently farts in yoga class.

John was guiding me so much, though, that I thought maybe I was having a placebo effect.

"You might be biasing me with all this priming," I said.

"I've primed plenty of people who have hated it anyway," he said, laughing.

It was a sad laugh, the laugh of the chronically underappreciated.

John drove me back to the bus stop. Maybe I saw something in the True Mirror and maybe I didn't, but either way, John helped shift my perspective. Trying to figure out what I look like by my image alone is like trying to figure out what a truffle tastes like by holding it in my hand. There's more in my face—in everyone's—than solely the superficial.

Faces are made real—unique and beautiful—not by what they look like, but by who animates them. No wonder the frozen frame of a photograph and hardened glances in the mirror—a place I hope to see beauty where I'm mostly searching for flaws—left me bewildered and inconclusive.

I stood in the parking lot, waiting for the bus to come and take me back home. Unlike in the city, there wasn't a constant barrage of reflective surfaces. There, at every turn, you are peering at, catching peripheral glimpses of, and judging yourself. In Rosendale, you are reflected by nature—the wind hitting against your skin, skimming and informing your curves, not trying to size them up or pin them down. A fly buzzed by my ear. A strand of hair tickled my nose. The sun warmed my cheeks. I was reminded, in those surroundings, of what it feels like to be inside a body, instead of what it feels like to look at one.

Mirror Face

When I look into a mirror, I contort my face—widened eyes, pursed lips, eyebrows arching upward into the nosebleed section of my forehead. I didn't realize I did this until I was with a friend, assessing myself before we went out to dinner. She said, "What the hell is going on with your face?"

I knew other women did "mirror face"—I'd seen it and thought it looked ridiculous—but I never thought I did, so I got defensive and said, "I don't know, what the hell is going on with *your* face?"

Then she playfully mocked me by doing Zoolander's Blue Steel model pose. I tried to laugh it off while wishing for the trillionth time since I was born that life had a rewind button.

Ultimately, my friend had a good point. Seriously, what the hell was going on with my face? How is it that a mirror compels me and countless other females to unconsciously vogue at ourselves?

For this, I spoke to Jennifer Davis, a sociology professor at James Madison University and a leading expert on the duck face, which often appears in selfies and is mirror face's closest pucker-lipped relative.

"Oh yes, mirror face," she said, like it was an old friend she'd recently gossiped with. She told me that when we do mirror face, we are contorting our features as close as possible to societal beauty standards. "You're creating a caricature of femininity," she said. "You are making your face more slender, your lips bigger, your cheekbones higher, and your wrinkles smaller."

I was impressed that I somehow knew how to do all of that. I don't even brush my hair.

"It's not calculated," she said. "Culture writes itself onto our bodies."

Davis suggested that mirror face has probably changed shape depending on the era, because it would align with what people considered attractive at that time. "Maybe women in the 1920s were seductively holding up cigarettes to their lips," she said.

The reason we do this bizarre behavior for ourselves—alone and in front of a mirror—is to rehearse what we want to look like to the outside world, as well as to create a sense of self that feels attractive and desirable. "You are saying,

continues →

'Okay, I want to look good,'" Davis explained. "'I want to control my face and body in ways that are flattering.'"

"But if we think we look hot like that, then why does it feel so embarrassing and shitty when someone calls us out?" I asked.

"Yeah, totally," she said. She explained that we devalue vanity in our society even though we necessitate it because of the value we place on beauty. "So it feels really vain, and that's part of the embarrassment factor," she said, "it suddenly feels really inauthentic."

In other words, our secret is exposed. We're ashamed because we've been caught performing being beautiful, when the world just expects us to be beautiful.

Davis had one last point. She explained that mirror face actually has a lot in common with duck face. When women are caught duck-facing, they may look attractive, yet they are often punished and mocked for putting too much effort into what we believe should be effortless. Instead of blaming the root causes—patriarchy, sexism, and misogyny—for creating this specific ideal of femininity, we blame the woman for showcasing it. "It's a catch-22," she said.

I had no idea that these facial contortions had such pernicious roots. I decided to end the patriarchy's influence over me, at least in this facet. Therein would lie liberation and self-acceptance. So the next time I gazed at my reflection, I stopped doing mirror face. Once you know you do it, it's easy to stop. All it took was consciously letting my facial muscles go slack.

Barely three seconds in, I could confirm that I wasn't into this type of liberation. My brows, once majestic and lifted, now looked like furry eye awnings. My lips were a thin and somewhat downward arching line. I no longer had cheekbones.

It was a conflicting sensation; I immediately wanted to let the patriarchy back into my life. Dammit. Mirror face was my jam. I tightened my muscles back up, breathed a sigh of relief, and went on with my day.

4

Earth Moved. Trust Me.

The first time I had sex, everything surprised me, but two things in particular: One was that condoms, if opened quickly, can shoot at your face like a taut rubber band, and the other was that those erotic moany sounds that I assumed came out of all women's mouths during sex were, for some reason, not coming out of mine.

I was sure that moany sex meant better sex; to me, the women who used those sounds while they humped were the poster children of the sexually free. I suspected that they were having the coitus of champions, which meant, of course, that there was something I was missing out on.

I hypothesized that as I became more and more comfortable with genital relations, rapturous orgasmic sounds would likewise develop. I fantasized that one sweet day, people would pass beneath my bedroom window, look at each other in both admiration and fear, unsure if the extraordinary din they heard from my humping was a result of ecstatic release or someone being stabbed multiple times.

So the second time I had sex, I was ready to get arrested for being a gloriously loud sex monster. We did all the same wonderful things and this time it was even more fun, yet when we finished, I was perplexed. I had once again gone full-on mime.

I knew that people faked, but that felt like cheating. I wanted to earn

my sounds. I expected them, when ripe, to come out natural
envisioned it, I wouldn't even be able to stop them; they'd be as
as it is for a dog to bark at an intruder (yet hopefully involve less

As the years went on, I had other boyfriends, but my decibel
mained disappointingly similar. By the time I was with Dave, my
landscape was still most closely matched with a library. Even th
we've discussed it and he said he doesn't need me to be vocally porn
can't help wondering if he's looking at me during sex and thinking, "Is th
thing on mute?"

So now here I am, a seasoned screwer, yet still as tight-lipped as I was in my awkward collegiate days. I've waited and waited, but these sexy sounds have refused to show up. To come to terms with my natural pull toward inaudible intercourse, I wanted to take a closer look at what I might be missing out on.

Are humans actually predisposed to being screamers, and if so, what is the purpose of all the noise?

CRIES LIKE THOSE OF DOVES, CUCKOOS, GREEN PIGEONS, PARROTS, BEES, MOORHENS, GEESE, DUCKS, AND QUAILS ARE IMPORTANT OPTIONS FOR USE IN MOANING.
—KAMA SUTRA

I knew it was possible that my ideal of a groany woman wasn't even rooted in reality, but rather due to the highly vocal women found in pornography. So before I went too gonzo, I wanted to know if sex noises were a new development or if they were an innate characteristic of *Homo sapiens*. Did our caveman ancestors set the mammoths stampeding when they bedded down?

To find out, I called Justin Garcia, an evolutionary biologist at the Kinsey Institute.

First, Garcia adjusted my terminology. "Sex noises" doesn't fly in academia. "Copulatory vocalizations," he corrected congenially. He then explained that for millions of years, humans have sexed it up in tight quarters—likely with other family members, children, and the threat of predators nearby. "So this idea that people, in the throes of passion, engaged in these really loud screams and uncontainable noises and vocalizations, it's probably not the case," he said. "Those vocalizations are probably part of the performative aspect of sex and less so about a natural reflex to sexual activity."

For a moment, I felt vindicated.

Clearly I am quiet because I have the lingering threat of a lion ripping my tits off mid-thrust imprinted into my DNA.

It is likely, Garcia continued, that porn has added a higher decibel level to our bedroom affairs. He explained that because we are a social species, we absorb the information around us and integrate it into our lives. "People observe shouting and moaning as the pinnacle of ecstasy and think, 'Here is a behavior that others are engaging in that they find arousing. I'm going to make it part of my sexual repertoire and I'm going to find it arousing.' That's part of the complexity of human behavior."

I suspected the same reason was behind why we shave our armpits, attach sparkly rocks to our fingers, and thought perms were a good idea.

Overall, I interpreted this as awesome news—for the first time in my life, my silence was making me feel like a radical nonconformist rather than uptight and repressed.

Before we hung up, though, Garcia warned me that there is a different school of thought, which posits that our sex sounds evolved for a specific purpose. He didn't want to go into detail because he didn't agree with the theory, but he told me that it had to do with hollering monkeys.

Before looking into the monkeys, I let myself revel in my quietness—or shall I say, in my deeply ingrained survival mechanisms. While all the loud chicks in my village would be panther food, I'd be quietly making babies near the fire pit.

When I was done gloating, I went to the library. I wanted to find out what Garcia was talking about.

My librarian deserves an award for not flinching when I asked, "Can you help me find this journal article—it's about the copulatory vocalizations of chacma baboons?"

I brought several studies home and read them while sitting at my kitchen table. It was basically very dryly written monkey erotica. Female baboons, it seems, make loud rhythmic sounds, which differ in length and volume depending on the ranking of the male the female is having sex with at that moment. These sounds, according to the dedicated researchers who watched the baboons hump for months near a swamp in Botswana, are used to advertise their sexual availability and ultimately, to attract males. This helps achieve the female baboon's admirable goal, which is to have sex with as many males as possible. By doing this, she will get a crack at making a baby with the best sperm out of the bunch as well as lessen the likelihood of a male killing her baby, because each of them will be duped into thinking they are the dad. This was interesting, but distant from New York City apartment bedrooms and the *Homo sapiens* having intercourse within them.

To interpret these findings and what they might mean for us, I got ahold of Christopher Ryan, the author of *Sex at Dawn,* a book that looks at the origins of human sexuality. He believes that, like baboons, humans are naturally promiscuous mammals and therefore would have developed moany, groany sex for similar purposes.

"So, are you saying that when a woman moans, she's doing it in the hopes of other dudes overhearing?" I asked.

"I wouldn't say that a woman who is making a lot of sound is consciously thinking, 'Hey, I want the guys in the street to get to hear this!'" Ryan explained, "but the origins of that behavior very likely involved attracting attention from other males."

Maybe that's why my husband has never complained about my lack of audible enthusiasm. Deep down, in a place he's not even aware of, he's pleased that I'm not attempting to incite our next door neighbor to come over and impregnate me.

"But what about the threat of predators?" I asked. I told him about Garcia's theory—that all these screamers would be dismembered mid-embrace by grizzly bears.

"In that case," he said, "why do we find female copulatory vocalizations in so many species of primates who are also dealing with predators?"

I didn't know what to say. Obviously, baboons have made it—they are alive and well despite their reckless yodeling.

This Christopher Ryan guy had a point. I didn't like that he had a point—he was taking away my newfound silent swagger—but I had to admit that he did have a point.

This was all fascinating, but even if it were the case, it didn't seem like the whole story. It was hard to imagine that all this screaming and hollering evolved solely for the attention of those outside the bedroom.

Even though Ryan wasn't very validating to my kind, I had an affinity toward him. As we chatted more, I mentioned another study I'd read with a small detail in it that had really tickled me. The study was about the sex sounds of yellow baboons. Researchers discovered that yellow baboons make the same loud hollers during copulation as they do when they are defecating—but really, the same exact ones.

"When they poop," I told Ryan, "it apparently sounds like they are having hot sex."

"That's interesting," he agreed.

Since I'd read that detail, I'd been mulling it over and would soon learn that it was another piece to the puzzle; monkeys aren't the only ones with that issue, after all. People also grunt in circumstances other than sex. People grunt when they lift a heavy box, open a tight jar, or go running at the gym. Many a Hollywood movie tricks us with sounds we think are from sex only to open the door on someone playing tug-of-war or struggling to reach something benign under a bed. We, like yellow baboons, even grunt when we go to the bathroom.

I talked to a friend, one who demanded anonymity, about that. "I really wish I didn't make that sound, but I do," she admitted.

Was it possible that the strained bathroom grunt shared the same origins with the titillating sex moan?

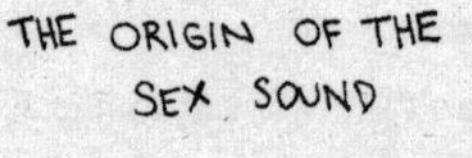

I spent days looking for someone who could speak knowledgeably about the genesis of grunting. Luckily, grunt experts exist and I actually found one to speak with.

Lorraine McCune has been studying the grunt at Rutgers University since 1987. She explained that the grunt is a physiological response to exertion, an epiphenomenon occurring when the body needs more oxygen.

What happens, more or less, is this: "Under conditions of metabolic demand, activation of the intercostal muscles to maintain lung inflation during expiration sets in motion reflex contraction of laryngeal muscles, creating a system under pressure that lengthens the expiration phase of the breath and enhances oxygenation of the blood. Expiration against the constricted glottis produces pulses of sound."

Translation: In the right circumstances, the sound just happens.

Now you can tell people the next time you get up from a sofa that you're not just being dramatic. The grunt is legit—you've got a tense glottis.

McCune went on to explain that tennis players often grunt when they hit a ball off their racket and that trying to stop the sound can actually hurt their game. "When you squash the grunt," she said, "you're having to use energy that you could have used for your stroke to suppress a vocalization."

There is even a study that proves McCune's point. Researchers from the University of Nebraska Omaha found that professional players increase the ball's velocity by 3.8 percent if they grunt while taking their shot.

When I read that, I got a little jealous—theoretically, during sex, the people who grunt enthusiastically can add force to their hump.

McCune's main focus is on how children develop language, the first step of which she believes is her beloved and oft overlooked grunt. Think of a preverbal toddler pointing longingly across the room at a ball: "ugh ugh." In light of this, she told me that she had about zero percent interest in discussing how this laryngeal sound—yes, verified to be the same one that often accompanies a strenuous bathroom session—can transform into a full-blown sex-sound storm.

This led me to Barry Komisaruk, a neuroscientist and the author of *The Science of Orgasm*, a man I knew would have no problem waxing poetic about these mechanisms. "No question," he said. "Sex sounds are a physiological response to exertion."

To tell me how sex sounds evolved from a small grunt into the screaming spectacle we know them to be today, he began by telling me a story about seagulls. "When a seagull begins to take off, it flaps its wings," he said. "Each time it flaps its wings, it makes a sound." He paused for dramatic effect. "Ahh ahh ahh," Komisaruk squawked, imitating the bird. "The vocalization is synchronized with the movement because the exertion creates sound."

This is where it got interesting: What begins as a simple squawking sound soon evolves to mean much more, he explained. A member of the seagull's flock that hears "ahh ahh ahh" will interpret it as a signal that his bird buddy is taking off.

"The sound serves as a type of communication, even though it wasn't the original intention," Komisaruk said.

The same goes for sex sounds, he explained. They may have begun as a series of small respiratory releases, but they have been adapted into a form of communication between partners. When a woman exhibits them, for example, they inform her partner about her level of pleasure and enjoyment.

"The sound is a representation of the intensity of excitation," Komisaruk explained. "If a partner gets excited hearing a shout during sex, then that can be a rewarding communication that bonds the partners and encourages them to do it again."

I've found that your lover can also be encouraged if you just take off your pants and awkwardly stare at him.

Meanwhile, I'd become intensely attuned to all the amorphous noises coming out of people's mouths. I was focused. The moan, in particular, caught my attention. It was similar to a grunt yet didn't require any prerequisite exertion.

Moans were in more places than just the bedroom. I was hearing them everywhere.

We moan when food is delicious, sometimes even before we take a bite. I myself got just about porny over a hamburger one night. We moan when we get a massage and the masseuse hits the perfect spot. We moan when we stretch our arms up, and we give a little "mmm" when a latte is just right. Why were we moaning like that?

I asked James Higham, an anthropology professor at New York University who specializes in communication, why we revert to amorphous moaning in these situations versus using the gift of articulate language, which we've developed almost miraculously over many millennia of painstaking evolution. In other words, why do pleasurable sensations make us go lexically Neanderthal?

In turn, Higham explained the law of brevity. The law of brevity states that the words we use most frequently are very short and the words we use rarely are long. "If every time we wanted to talk, 'yes' was replaced by 'sesquipedalianism,' then our sentences would be absurd," he said.

The briefest and easiest form of communication, he explained, of course, is a sound.

"I don't have to say, 'Oh yeah, that's the spot right there, no wait, just slightly up,'" he explained. "I can just be quiet until they hit the spot and go 'mmm,' and there you go—they know."

Higham gave about twenty-eight examples about when moans could be helpful.

"If your mom makes you lasagna and you moan," he said, "she'll know to make it frequently."

There might be a lesson here: If you want something to happen again, punctuate the activity with a vague low-range vocal hum.

After talking to Higham, I discussed the theory with a friend. During our conversation, she realized that she used the law of brevity during sex, too. Moans were a form of GPS for her lover. If she becomes quiet, then he knows he's lost his way. As long as he continues to follow her moans, he'll reach her desired destination. Essentially, they were playing a game of hot and cold with her genitals, except that in place of "hot" she used various copulatory vocalizations.

"Totally works!" she said.

The moan, then, was not only an exaggerated physiological reaction or an antiquated way to get attention, but also a shortcut—a way to be efficient. The moan, that little mush bucket of stretched-out vowels, started to seem even mightier than I'd given it credit for.

Another way to understand the significance of the sex sound was to investigate why women faked. I had never thought about it this way before, but women wouldn't go through all the trouble to put on such a performance if these sounds didn't wield significant power and influence.

Gayle Brewer, a professor of psychology at the University of Central Lancashire, coauthored a study about fake sex sounds. In the unimaginative yet fittingly titled study "Evidence to Suggest That Copulatory Vocalizations in Women Are Not a Reflexive Consequence of Orgasm," Brewer found that all her seventy-one respondents faked some of the time, while 80 percent of the women faked 50 percent of the time.

"They were doing it quite a lot," she said.

She found that women tend to fake for two different reasons.

One was that they wanted the sex to end. They were over it and ready to move on. This could be because they were pressed for time, tired, rubbed raw, or bored.

"It can become very abrasive," Brewer said of sex.

Brewer explained the method tends to work, too. Because sex sounds give a signal to a woman's partner that she's had her orgasm—which she usually hasn't if it's during intercourse, because, let's admit it, the clitoris is about a two-hour drive from the point of penetration—he feels like he can go ahead and let 'er rip.

"By vocalizing," explained Brewer, "women are saying, 'It's okay, you can go ahead and finish now.'"

So faking a state of euphoria, as counterintuitive as it sounds, is actually a polite way of communicating "Get the fuck off me now!"

The other reason women tend to fake is that they want to give an ego boost to their partner. The sounds act as an audible pat on the back, an enthusiastic thumbs-up.

"They are trying to reassure him that they are satisfied," Brewer explained, "and positively affect his self-esteem."

I wish I'd read this study earlier in my life. I've always gone with the theory that giving your lover the silent poker face makes him work harder.

The researchers posit that by boosting the man's self-esteem, he'll be more likely to come back for seconds. Seconds might turn into thirds. The more sex, the more chances there will be for a condom to break and thereby aid in the continuation of our species. (In the study, Brewer worded that differently, though she made the same general point: If a woman and man have a lot of sex, they will be more likely to procreate.) I would bet a lot of vocal women didn't realize that their vocal chords were trying to get them pregnant.

Ultimately, I felt mixed about all this news. In one sense, it made lady sex noises seem inauthentic. In another, it made them brilliantly strategic—like over the millennia women have expertly harnessed their vocal chords and turned them into a type of superpower. Using fake sex sounds should come with a cape and a leotard. With her voice, a woman can make a guy fall in love and/or ejaculate on command.

It seemed like sex sounds weren't the only thing that evolved—using them strategically did as well. Maybe these sounds deserve a little more respect. Instead of calling them fake, maybe we should make them sound a little classier and call them faux.

But no matter what we call the sounds, the message from Brewer's study is clear: A lot of moaning occurs when there isn't much to moan about after all.

The only question I had left, and maybe most important to me of all, was whether or not I was actually missing out on something by not being vocal. Did sound do more than just communicate and in fact actually enhance the sex experience? In other words, when I'm eighty and tucking in my grandchildren for the night, will I be tempted to warn them about the mistakes in my life? "Dearies, I only have one regret: Grandma should have fucked louder."

I found experts—people that I suspected would have intelligent input.

Nan Wise didn't hesitate a moment. She thought I was missing out big time—said it was a scientific fact. Wise is a neuroscientist at Rutgers who studies the female orgasm and also moonlights as a sex therapist. She told me about this thing she called the "throat-to-pussy connection."

"When you gag, like say you're giving a blow job and you gag," she said, "sometimes you will feel a contraction in your vagina."

I hadn't had that experience exactly, but she assured me that it's a thing that happens with regularity. She said this connection is also common knowledge in birthing classes where coaches will often tell moms-to-be that a relaxed throat will also make for a relaxed vaginal canal. "Everything is interconnected," said Wise. "When you activate the throat during sex, it can actually be very pleasure enhancing."

"How about when people talk during sex?" I asked. "Does that help?"

Wise said that talking was a whole different beast and not necessarily a pleasure-enhancing one. "Some people are just chatty motherfuckers," she said.

Then there was Barbara Carrellas, a sex educator and the author of *Urban Tantra*, who was on the same wavelength as Wise. "Obviously no one has ever died from not making sex sounds," Carrellas assured me, "but they add so much to the erotic experience." She explained that if you're not making sounds, then you're probably not breathing very much, and breath is absolutely critical for an expanded orgasmic experience. "All sex is about energy and sound brings energy," she said. "I mean that in the physics sense, not in the woo-woo sense."

I'd always felt that I should make sounds only if they were so powerful that they could not otherwise be stopped—that's the only way I felt that my sex sounds would be authentic. The physical sensations, in my mind, had to lead to the audible.

Wise and Carrellas were saying it was okay and even good to try the opposite. Sometimes it was the vocalization that could actually drive, augment, and incite the physical response.

I continued to speak with an inordinate number of people—voice teachers, prenatal yoga coaches, sound healers, Taoist gurus, and even a group of women who call themselves sensualists; they spend a large part of each day having orgasms. It was unanimous. Every single one of them lauded the sex sound. They weren't advising to do it to please a partner or to playact sexy and over-the-top like a sexpot porn star; they believed that our own unique sound—whatever that may be for each one of us—could legitimately expand our own pleasure.

Donna Reid, a voice teacher for the past twenty years, even got metaphysical: "When you free the voice, you free the self."

The feedback was becoming quite compelling; it sounded like it paid off to be a vocal woman.

Over the next few days, I digested all this information, and some surprising emotions emerged. Even though being loud had obvious advantages and was something that I'd aspired to since a young age, I began to feel righteous about my place in the vocal continuum. I felt like one of the little guys who must stand up stoically for a different way of life, like Amish people. No matter how silly their buns might look covered in their little bun caps, they do it because, gosh darn it, that's who they are. I even thought about starting a silent-sex chat room in order to give support to other silent sexers all over the world. We could band together and petition for our kind to be represented in Hollywood films.

But then one evening my curiosity got the better of me—I decided to do it like a baboon.

I'd warned Dave that things might be different, but he wasn't prepared for what happened that night. He laughed a lot. I laughed, too. It was uncomfortable. I sounded a bit like Pee-wee Herman trying to use a toothpick to till a large garden. No one told me that it might take a while to find my sex voice, but as with most art forms, I think that it's true.

Back when I'd spoken to Wise, I'd asked her why a woman might not make a lot of sound. First, she theorized that this woman might be repressed, but then she said something else I found quite insightful. "Maybe she wants to concentrate on her own sensations," she'd said. "It can be a way to be focused on the inside, on what's going on for you."

I liked that reason and it resonated.

When people go blind, their other senses often pick up the slack, leaving them, for example, with super ultrasonic bat levels of hearing. Maybe if you aren't busy hollering during sex, you have the space in your brain to develop advanced sensory sensitivity in your vagina.

I'm not saying I'm not a little repressed as well. I'm obviously a little repressed. I also refuse to do karaoke or let loose on a dance floor without giving myself alcohol poisoning first. I guess what I'm saying is that I'll keep experimenting with sound—I will—but at my core, I'm a silent fucker for life.

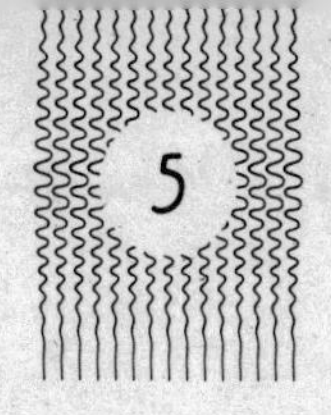

The Big Dripper

During much of my adult life, it has not been uncommon for random people on the New York City streets to ask me, out of the blue, if I had a good run. I look around confused—I'd just walked to the mailbox or grabbed some lunch—until I realize that they are staring at, and trying to find context for, the beads of sweat swimming down my forehead. I give in. I don't want to have to explain that actually I'm just disgusting.

"Yes, great run," I lie.

Throughout my life, sweat has caused me numerous moments of discomfort. There was that job interview: I finished it up feeling great about myself, only to look down and discover that I'd had armpit sweat spots the whole time. I looked like a paper towel commercial, but instead of a paper towel, it was my shirt that was touting its absorbency after a spill. Two circular pit lakes had combined right in the center of my chest to create one massive heart-shaped spot of shame.

I did not hear back about the job.

My wedding was another sweaty situation. While most brides are concerned about DJs and flower arrangements, my biggest concern in the weeks before the big day was how I could be sure that my white dress would not look like I had won a wet T-shirt contest. To ensure dryness, I used a product that will probably give my child, if I have one, eleven fingers.

And yes, it was worth it.

I've learned how to hide behind tree trunks and jam my hand down the top of my shirt with a napkin to take a swipe at drying my mid-chest boob sweat without anyone seeing. Before getting dressed, I first imagine what my outfit would look like with wet splotches. My wardrobe, because of this, is mostly made up of black. In a very self-defeating practice, I've been known to keep my jacket on, despite being hot, just to cover whatever mess my skin has already made inside. Of course, things only get worse. Of course, I never learn. At least I no longer make the mistake of trying to dry off my sweat with a bathroom hand dryer. Never, ever do that. Not ever. Any drying progress you make is ruined almost immediately by the intense heat of the dryer, which only makes you sweatier, coupled with the anxiety sweat that occurs as you consider the possibility of an acquaintance imminently walking in on your precarious position.

I could go on and on about the episodes I've experienced and tactics I've used to camouflage or downplay my sweat, but I'm more curious about looking into why sweat feels like such a big deal. Why, depending on context, is it so horribly embarrassing? We are built to sweat—it's not only normal, but actually essential for our survival—so why do we spend a collective eighteen billion dollars trying to pretend that we don't have pores? I'm not looking to find a way to stop perspiring or even to overcome the shame, I only want to know why I can't sweat (other than inside a gym or a sauna) and have it not be weird.

I can only wear black

I thought I could answer this question by approaching Lillian Glass, a human behavior and body language expert. She also acts as an expert witness in court cases—evaluating the comportment of those being tried—and advises people on how to hide their tells. For those who feel that they blush too easily, she'll train them to apply green-toned foundation to neutralize the revealing red.

I wanted to know, from her point of view, what sweat communicated to those around us. Why, during a barbecue this summer, was I compelled to go to the bathroom to stuff my armpits with toilet paper and replace the wad every half hour? Rather than show my sweat, I probably left the festivities with the attendees thinking I had bulimia or misplaced implants. Did I actually have reason to be so afraid of some measly wet marks?

Before I got on the phone with Glass, I wondered if I'd been blowing all this sweat stuff out of proportion—been fixated, obsessed by perspiration. In all likelihood, Glass would laugh at me and say, "Honey, sweat is as natural as eating a whole pint of ice cream alone in the dark, so go find yourself something worth your while to be concerned about."

But in less than a minute, it was clear that the conversation would be veering in another direction. Glass, a professional accustomed to the buttoned-up atmosphere of a courtroom, couldn't even keep her disgust in check. "Sweat is icky," she said. She took a pause before what I thought was going to be an elaboration, but I was wrong; she just wanted to thrust the knife in a little deeper. "Yes, there is a definite ick factor," she reiterated.

"But why?" I asked.

She said in a society that often likes to mask emotions, sweating is a tell—it shows what's really going on inside. She said we sweat when we are sick, anxious, and deceitful. She knows when a defendant is lying or nervous because of an instantaneous beading on the upper lip and forehead. Sweating, she said, can also connote bad health, obesity, and laziness. "You aren't keeping your cool, so to speak," she said. "Something is taking place inside of your body that is not attractive." Also, the mere visual of sweat begets the fear of a looming odor, she explained. "We are having a biological reaction to the sweater's biological reaction."

"But what if it's just hot outside?" I asked. I've got glands that seem set to spurt on a hair trigger. It's not fair. Surely it's not evil to sweat when the sun is bearing down.

"A drippy, sweaty face is not attractive," she reiterated.

Maybe it was the fault of movies. Sweaty roles are reserved for characters who are suffering from food poisoning, are about to turn into a zombie, or are caught with child porn. You never see the beautiful ingénue sporting a pair of sweat rings just for kicks. According to Hollywood, the only time good people sweat is when they are at the gym or in the midst of mind-boggling sex (which is confusing because, depending on context, the same bland and innocuous liquid can either be associated with shame and embarrassment or hard work and the erotic arts). This paradigm doesn't give sweaters like me a lot of leeway to be cool while also being hot.

Glass kept telling me how to fix it. "You need to mask it or control it because it can be uncomfortable for the person looking at you."

I didn't want to believe this, but I knew that this was kind of true. I've shown up to a party with droplets suspended on my temples and a glimmer of underboob sweat, and people have gazed at me with the same mix of concern and disconcertment I'd expect if they'd just caught me fornicating with an antelope.

"There are many things you can do!" Glass told me to put ice on my wrists. She said to dab my face with powder.

A study conducted by Harris Interactive, a market research agency, found that roughly half the people interviewed were more embarrassed by extreme sweating in public than they were of having their fly down or passing gas.

"I guess part of me wants sweat acceptance," I finally told her.

"I wish I could say that we would accept you . . . " Glass said.

Could a sentence begin more ominously?

"But the reality is that we're not going to." She told me once again that sweat makes people unattractive and that even celebs get lambasted if they are caught with their glands leaking. "People aren't supposed to be wet, otherwise you'd be a fish."

"A fish," I echoed.

"I'm sorry I don't have better news," she said.

There had to be a better or at least more diplomatic way to examine this function. Glass obviously had no sympathy for sweaty beasts. But I decided to take a step back and first learn more about the divisive liquid. Was there anything, physiologically, that made it deserving of its odious reputation?

Nigel Taylor, a professor at the University of Wollongong in Australia whose passion is human temperature regulation, agreed to speak with me. He applies his knowledge of perspiration, which is deep and expansive, to problems such as figuring out how long soldiers can partake in combat under the weight of their gear without getting heatstroke.

He gave me the basics of sweating, none of which lead me to believe heavy sweaters should be social outcasts. We have thermoreceptors all over our body—the chest and spinal cord among other locales—which report to the hypothalamus. If we are exercising or find ourselves outside on a scorching summer day, the thermoreceptors will deliver an important message. "They say, 'Shit, it's getting hot in here. You need to do something about it now!'" Taylor explained.

First, our blood vessels will dilate, bringing them closer to the skin's surface, which begins the process of releasing heat. If we do that and are still too hot, that's when we begin to sweat.

Taylor continued to explain that we have two types of sweat glands—apocrine and eccrine glands. Apocrine glands are mostly in our armpits and groin and tend to leak during more emotional moments like trying to dig up the perfect change to buy some K-Y Jelly while a line of people behind you impatiently stare. Those glands produce viscous, oily, and slightly opaque sweat. Bacteria love to feed on this type of sweat and can leave an off-putting odor (the ideal complement to those stressful moments!). Eccrine glands, on the other hand, are located all over our body—there are millions upon millions of them—and they expel what is essentially just water with some traces of sodium.

In a perfectly efficient world, we would sweat, but you'd never see any evidence of it—we'd sweat only as much as could be immediately evaporated from the skin (which is the actual process that cools us down)—but many of us don't work that way. The wetness we see on our skin's surface is excess; it's like pouring car coolant out of a bottle and onto the ground. Taylor told me that it's mostly useless except for making us feel awkward, but there is clearly also another way to look at all that extra liquid: Our bodies are being overachievers. We are winning so hard at the cool-down game.

Besides, too much has to be better than the opposite, right?

"So what would happen if we couldn't sweat at all?"

"If you can't rid yourself of heat," Taylor said, "you'll die of hyperthermia in twenty minutes."

Hyperthermia is, as the name would imply, the opposite of hypothermia; it occurs when our body temperature gets too high. The condition causes tissues to break down and organs to fail. "When that happens, that's it for us," he said. "We're dead."

I liked that answer. Sweat has gravitas; it is the beating heart of our skin. We need it to go on and on and on.

Sweating, as Taylor explained, is not only entirely and utterly normal, but also necessary, so I wanted to know what gave the liquid such a toxic societal association. I began looking to the past. Something must have happened during sweat's upbringing that caused it to turn from an everyday Joe into a villain. I felt like a therapist trying to figure out why my patient began to enjoy dismantling gophers. Something had happened during his formative years, but what?

It took some time, but while searching through old medical journals and studies, I discovered a German medical historian and doctor, Michael Stolberg, who'd spent more than a year trying to untangle our ancestors' views on sweat.

"I was surprised," he told me later. "When I embarked on this, practically nothing had been done on the history of sweat! Nothing!"

Stolberg also gets excited about uroscopy, the practice of diagnosing an illness based on the color, smell, and even taste of a patient's urine. (He actually wrote a whole book on this.) "In the 1600s, they had the most beautiful and detailed paintings of urine," he told me. "You should check them out. The colors refracting. Just beautiful."

Officially, Stolberg one-upped me in gross.

To piece together a coherent picture of early sweat perception, Stolberg translated old textbooks, medical records scribbled in Latin, and letters written by people who were trying to make sense of their sweat-soaked inheritance.

He focused on a three-hundred-year period—1500 to 1800—and found that sweat, for people back then, made for a precarious catch-22. "On the one hand," he said, "sweating was very welcome. You needed to sweat in order to purify your body—it had a cleansing effect. But on the other hand, the stuff that came out of you was gross and people had a fear that this substance, even in vapor, might reach someone else and make them sick."

"Sounds tricky," I said.

"Absolutely," he agreed.

To recap this dilemma: It was essential and healthy to sweat; yet that same sweat was considered by others who might come into contact with it as being as dangerous to one's well-being as standing before a hungry and rabid lioness. (Okay, not quite that dangerous, but still dangerous.)

Stolberg went on to tell me that at the time, physicians and laypeople alike thought that clogged sweat glands were among the leading causes of illness and death. To illustrate, he recounted a tale that he'd documented in his 2012 sweat treatise "Sweat: Learned Concepts and Popular Perceptions 1500–1800." A man in the 1700s found his sweaty feet a pain in the ass, so he began slathering fat on them during walks. The amount of sweat decreased and almost simultaneously his sight began to grow weak. "The way they understood it," Stolberg said, "was that the harmful matter, which, until then, had been excreted from the feet, had now turned toward the eyes."

In another case, two hundred years earlier, a man named Achatius Trotzberg sought a doctor's help because he'd started to experience pain in his stomach and limbs. He told the doctor that he suspected it was because the once nearly constant sweat storm stemming from his legs and feet had recently fallen to a measly drip. The lack of sweat clearly meant that poor Achatius couldn't get the bad stuff out, hence his body, with all the bad stuff stuck inside, was turning on him.

I wish Achatius was correct, because if he were and sweat got all the bad stuff out, then given my propensity for leaking, I'd live until at least 145.

People were so afraid to screw with their sweat output that they wouldn't use cold water, lest it close up their pores, hence blockade perspiration and render them unwell. The misunderstanding was easy to make. For one, doctors had observed ill patients who had suddenly felt better after a big sweat. Instead of interpreting it for what it was—a sign that a fever had broken—they thought the sweat itself had carried the illness outside the body.

Then Stolberg hit me with this gem: "It's kind of funny because people still believe some of this today," he said, "that sweat cleanses or detoxifies the body, even though there is absolutely no science to back it up."

I would have thought it was funny, too, except I suddenly realized that I might be one of those people he was talking about. "Sweat isn't cleansing?" I asked. "Like at all?"

I mean, I knew I wasn't going to catch a disease from the stuff or die if I wasn't juicing from every pore, but I'd definitely had my fair share of sauna sessions after drinking too much alcohol, suspecting that somehow the sweat would clear me of my hangover. I'd also religiously attended Bikram hot yoga for a year and could swear I'd felt detoxified with a renewed sense of sentience after peeling off my soaked clothes. (I also loved being in a place where the dry people were the marginalized ones.) I funded my habit by volunteering to wash all the sopping-wet towels that the paying clients left behind. It was the most disgusting task I've ever been given and that's even when I factor in that I've popped a friend's butt pimple before.

"Of course sweat doesn't cleanse the body," Stolberg said, laughing.

This was such a crazy revelation that I felt this new-to-me, yet old-to-Stolberg fact needed additional validation. The only reason I felt comfortable reaching out to ask more professionals, was that when I'd told friends—really smart people who can make their own beds and have been to college and everything—they, too, had thought that sweat had cleansing effects.

"You got to be fucking with me," one said, when I told her that apparently hot yoga is really only good for dehydration.

All the people I asked—a dermatologist, a physician, and a physiologist—chortled under their breath before reiterating a similar refrain. "There might be trace elements of something like garlic," said David Pariser, a dermatologist on the board of the International Hyperhidrosis Society, "but toxins are removed through urine and feces. The liver is what detoxifies." There was enough condescension in his tone to humble a pack of arrogant Wall Street lenders.

If that myth has stuck with us through all these centuries, then maybe, in a more subtle way, the early belief that sweat is a dirty and contagious liquid has as well.

But what probably led to the greatest alienation of the sweaty amongst us occurred in the early twentieth century: advertising! Or more specifically, the advent of antiperspirants.

I'm keeping this part short, because it closely echoes the same trajectory and promises that the invention of razors (and the ensuing advertisements) had on female body hair: salvation and sexiness via complete eradication. Basically, advertisers like to create insecurities—bodily ones seem to be their favorite—and then exploit them ruthlessly. Women's armpit sweat was first targeted for annihilation while men's sweat, for at least another couple of decades, was still viewed positively. A sweaty man was masculine. Wet pits showed how hard he'd worked out in the fields. Women's sweat, meanwhile, became a daily atrocity.

At first, women wouldn't play along; they were tentative about clogging up their glands. They still believed dammed sweat could cause an internal disaster, but advertisers in the early 1900s persevered and upped the ante. They not only tried to educate women—"no harm will come from stopping the troublesome perspiration in *limited sections of the body*," read one ad from the era—but also aimed to convince them that sweat causes severe shame and a botched sex life. I'd say they were successful, as those are both feelings I have experienced myself.

Have I told you the one about the sweaty girl who walks into a bar to meet her internet date? No? Good. That will be for me and my deathbed.

Advertisements for popular early antiperspirant brands such as Mum and Odo-ro-no (odor? Oh no!—so clever) depicted such sad scenes as a gussied-up gal with a solemn face and nowhere to go because, you guessed it, her pits were rank! The ad is accompanied by this copy: "You're a pretty girl, Mary, and you're smart about most things. But you're just a bit stupid about yourself . . . You've met several grand men who seemed interested at first. They took you out once—*and that was that*. WAKE UP, MARY!"

There are umpteen of these ads, but I'll just drop another treasure right here:

> *If you think you don't perspire enough to matter, just smell the arm-hole of the dress you are wearing when you take it off tonight.* It *won't hesitate to tell you that you're no exception to the rule! You'll know at last why romance is passing you by!*

After one hundred years of relentlessly informing us about how disgusting we all are, the message seems to have taken a firm hold. Visible sweat is so inadvisable and unexpected—unless you're wearing yoga pants and a Lululemon sports bra as you walk out of SoulCycle and inscribe #soulsweat on your Instagram photo—that if it happens to you, you must have an excuse ready.

"Sorry, I'm super gross" works for me sometimes. You're welcome.

Sweating, officially, has become more than a body function; it's now yet another problem to be fixed. I observed this firsthand when I talked to Michael Brier, the owner of Kleinert's, a company that has been selling "sweating and odor solutions" since 1869. They make pads that look like disposable diapers for overactive armpits and swabs doused with aluminum chloride that are supposed to stop your skin from leaking for seven straight days. I thought it would be enlightening to get the viewpoint of someone who'd made it his life's business to thwart sweat.

"I'm not embarrassed or ashamed about what I do," he said in response to my questioning. "I feel actually that I'm doing good for society. I'm helping people, like yourself, and I don't think there is anything wrong with that. I'm not a murderer."

I may have, not purposefully, put him on the defensive. I couldn't help it; I viewed him as an unwitting villain. He was normalizing the ouster of sweat from our everyday lives. He was why everyone with sweat glands is expected to be dry.

"You're a sweat murderer," I said.

"Okay, well sure, that's one way to put it."

I went on to ask him more about the people who depend on his products.

"Why is there such a demand?" asked Brier. "Because there's a problem. Why is there a problem? It's just the way God made us."

If there is a God, and we're getting hypothetical now, it's doubtful that the all-powerful one would invent something just so that us humans could come along and obstruct it with a chemical compound and cotton padding.

In order to give sweat the respect I felt it deserved, I clearly had to go back to the very beginning. I had to put sweat back into its most original context—before the antiperspirant ads and the Renaissance doctors. I had to show sweat at its most innocent and vulnerable, when it was a newborn and free from any connotation. To do this, I contacted Daniel Lieberman, a paleoanthropologist and the resident sweat theorist at Harvard University. He has spent his career trying to figure out why the body functions the way that it does.

I started out with the simplest of questions: "So why did we start to sweat?"

Lieberman explained that for many millions of years, mammals have been sweating, but mostly from their paws. "When we get nervous, we sweat on our palms, too," he said. "That's part of our ancient heritage for escaping."

He explained that the wetness created traction so that animals, when in danger of being eaten, could climb up a tree or scale a cliff. "Think about how you lick your finger when you want to turn a page in a book," he said of how the traction works.

This information caused me to revisit the question: If you could go back in time, where would you go? I used to think I'd go back to witness Michael Jackson's Victory Tour, but now I'd go back to when palm sweat was for champions. These days, sweaty palms have no cachet. After shaking hands with someone, I always have to pretend that I just went to the bathroom and care about the environment. "I hate wasting paper towels," I say as my victims try to nonchalantly dry their hand on the side of their pants. If it were three million years ago, no one would be grossed out; they'd call me a survivor.

Lieberman told me that this kind of sweating—the clammy-hand kind—originally had nothing to do with regulating our temperature.

"So we sweated from nervousness before we sweated to cool down?" I asked.

"Millions and millions and millions of years before," he said.

That must be why I'm so phenomenal at nervous sweat; when giving a public talk, I've been known to leak through a quarter-inch-thick sweatshirt.

Way back then, to stay cool, we would have panted like dogs and monkeys. Then at some point, though no one knows exactly when, our internal air-conditioning model shifted: We stopped panting, lost our fur, and became much more sweaty.

Lieberman has spent many years considering why this occurred—possibly even as many as I've spent considering how to hide underarm sweat while wearing a long-sleeved button-down denim shirt, but luckily, he's been more successful. While I have two such shirts, still with tags on, hanging in my closet, he's come up with two viable theories.

In one, he noted that when we began walking upright—on only two legs instead of the customary four—we also became a lot slower. "It was dangerous to be in open habitats," he said. "We were vulnerable." To survive, he believes we evolved the ability to sweat so that we could forage at times of day when other animals couldn't. "Lions, hyenas, and all the predators that like to kill can't be active when it's very hot," he said. Sweating, in other words, gave us a niche. We could gather food in peace because our panting predators couldn't handle the midday heat.

His other hypothesis was much more specific. He suggested that thermoregulatory sweat began two million years ago, at the same time that we began endurance running. "When you run, you generate much more heat," he said, "so you need to have the ability to dump that heat much more effectively."

Because sweating allowed us to run great lengths in hot temperatures—a unique trait matched by very few species, such as horses—our Paleoproterozoic grandparents back in Africa were able to do something called persistence hunting (it's such an important development that anthropologists often refer to it by just its initials, PH, like it's a famous rapper or a popular drug that makes people eat their own faces). What would happen is that during the hottest time of day, early humans tracked and chased big bloodthirsty beasts. These animals, though powerful, weren't built to do something as simple as pant and gallop at the same time, so they couldn't cool down, and therefore tender, scrawny, perspiring meat bags such as ourselves were able to run these animals to death. After overheating, they'd become our meal.

Lieberman kept talking, but I wasn't listening anymore. This was it! Sweating, as he'd just noted, was a tremendous progression for humankind. Because of sweat, we were able to hunt big prey, which means that for the first time, we regularly ate meat. As a result, our teeth got smaller, as did our stomachs, while our brains got huge. If it weren't for sweat, I wouldn't be writing this right now. I'd be crouched in a field, using a rat's tail as a thong while ingesting bark with my caveman sweetheart, Oog.

"I get it," I said. "Without sweat, we wouldn't be the intelligent, big-brained social animals that we are today."

I was reveling in my revelation—pursing my lips and nodding my head with pride. If there were a flag for humanity, it should be of a face with a sweat droplet rolling down its temple.

But Lieberman, I soon realized, was a man who did not appreciate joy. He was not at all down with my proposition. "Sweat helped us become who we are," he said, "but we didn't have to be the way we are."

"But if we weren't us, then what would we be?"

"If we hadn't evolved to be the way we are," he said, "then we would have evolved to be something else."

"But then we wouldn't be us, right?"

"We would be us, but a different us."

Maybe he was saying we would have grown air vents on our behinds or sprouted five-speed fans from our shoulders. Whatever it was, it didn't happen. We do sweat and I finally had perspective. "Pit stains aren't repulsive," I said, trying to get him onboard, "they are our legacy."

Showing up to a meeting with sweat blotches on a fresh blouse and an upper lip dense with water beads is actually a symbol of our fortitude as humans and a gift that binds us to our prehistoric ancestors. When we look upon a hot mess of a woman, we should bow down to her shiny complexion, for it is that sheen of liquid that brought us steaks, feather mattresses, and even the internet.

"There are so many components to our bodies that to ever say that one thing is what makes us the way we are is just facile," Lieberman said.

"But—"

"Would we be the way we are if we didn't have feet? Language? Cooking?" He told me that everything is contingent in evolution, so you can't focus on just one aspect. "This is very sloppy thinking," he said.

This Harvard professor's condemnation of my critical-thinking process made me start to spout water from my more occluded regions. By the time we got off the phone, I'd built up enough traction to shimmy up the trunk of a palm tree. It was going to be hard to rebrand sweat as the pride and joy of our species if Lieberman wouldn't back me up.

I went back and thought about sweat long and hard, so long and hard that I saw a breach in my logic. I'd preached for people to embrace sweat, yet if I'm being honest, I have to say that I myself have difficulty when I see someone gushing in public. I thought about a meal I once had in the Mexico City airport. My waiter at the Chili's outpost sweated profusely from his head. I felt concern—was he going to have a heart attack or drip into my food? Neither option sounded good. After all this talk of acceptance, I had to recognize that I was just like the homophobic male senator who is secretly having an affair with his male page—a total hypocrite. Maybe there was something deeper to the aversion. Could it really just be the fear of creating an odor and the unsavory connotations attached to sweat—illness, laziness, deceitfulness—that repulse us, or is there something more complex going on?

FLAG OF HUMANITY

For this, I had to move away from the sciences and engage with some humanities. I got in touch with social psychologist Jamie Goldenberg, a professor at the University of Southern Florida. I reached out to her because her research focuses on people's relationship to their own and other's bodily functions—all the good stuff like sex, poop, sweat, menstruation, and vomiting.

"So from your perspective," I asked her, "why do we spend so much time and money trying to hide our sweat?"

Goldenberg told me about terror management theory, which looks at how we deal, often unconsciously, with our existential fears. She brought up Ernest Becker, the philosopher responsible for this line of inquiry. "His most famous quote," she told me, "is 'We are gods with anuses.'"

What a charming expression. I'm not sure why I haven't seen it inscribed in cursive on a Hallmark card before.

"We want to be gods," she explained of the quote, "and put ourselves on a different plane from all other creatures, but the anus is a constant reminder of our creatureliness and our eventual death."

That resonated with me—my anus continuously disappoints me (I'll get to that later)—but I wasn't sure it related to the subject at hand. "So what does that have to do with trying to pretend we're not sweaty beasts?" I asked.

"It has everything to do with it," she said. "See, we don't want to be beasts."

She explained that those parts of us that are animal-like—snotty snouts, pus-filled boils, milk-laden mammaries—are reminders, often subconscious, that just like other animals, we have an expiration date.

"So we invest a lot of effort in our bodies and try to transform them from their natural animal-like ways," she said, noting practices such as wearing makeup, clothes, tattoos, and perfumes and even engaging in practices as simple as brushing one's hair. "It's a defense against the anxiety associated with the awareness of mortality."

Goldenberg's theory sounded plausible, but I had my doubts. "Okay, but how do we know that this is all stemming from existential angst rather than a persuasive advertising campaign?"

"Why do you think people are so receptive to these types of advertisements?" she said. "I would suggest that it's because they tap into our basic fears and yearning for self-regard. They remind us that we have physical weaknesses, the biggest of which is that we'll eventually die."

Coincidentally, I've also always considered dying a significant weakness. But suddenly, I could see why sweat was so disconcerting. We live in a world where we try to control everything, but sweat doesn't cooperate with the mind. Sweat is a reminder that instead of us ruling the body, the body rules us—with each rivulet, we see that we are tethered to a time bomb.

To go a little deeper, I spoke to one more academic, Sheldon Solomon. He cowrote the book *The Worm at the Core: On the Role of Death in Life* and is a psychology professor at Skidmore. I had been hoping for sweat acceptance, but after hearing about why we might be repelled by our own orifices, I wondered what the fallout might be if we dropped all our pretenses and were able to embrace ourselves as the animals that we are.

"Would we all be better off if we could be more in touch with our sweat and maybe therefore also our own mortality?" I asked Solomon.

"I think yes and no," he said. "Certainly I think it would not be a bad idea to realize that we have almost turned looking and smelling good into a fetish."

We both agreed, being prolific sweaters ourselves, that it would be nice if people didn't have an expectation of dryness, especially on hot, humid days.

"On the other hand," he said, "I don't think it would work out well if the only thing we did was walk around thinking, 'Oh my God, I'm a breathing piece of defecating, decaying meat.'"

A few days later, I was in the kitchen, absentmindedly rearranging magnets on the refrigerator while trying to make sense of everything I'd learned. With all the information, my mind was only getting more scrambled, so I went to bother Dave.

The poor chap was working at his desk when I barged in. He swiveled his chair around. "Can this wait?" he asked.

I sat down, ignoring what he'd just said. "It's about sweat."

Dave knows that I have issues with the stuff, that not only am I a profuse sweater, but that while I try to accept it and embrace it—oftentimes by refusing to use antiperspirant—I nonetheless get uptight and embarrassed when I spring a leak in public. It's a deeply contradictory way of being that I don't recommend to anyone.

"I get annoyed that we are expected to hide these very natural things," I said, scooting deeper into the chair.

Then Goldenberg, the psychologist I'd recently spoken with, came to mind. "And yet, maybe hiding it helps us not think about mortality." A couple of seconds later: "Did you know that sweat doesn't release toxins?"

"Get it all out," he said.

"But I do agree that seeing someone sweat in some contexts can make people uncomfortable," I said. "Remember that waiter in Mexico?"

He nodded.

"Yet advertisements try to make us feel like crap all the time just for being normal." I'd been wondering if advertisers began a campaign against yawning, how long it would take before we'd all begin to buy specialized masks, produced with the specific purpose to hide the involuntary reflex.

"Sweat is amazing, though. It helped us hunt big animals," I said. "It's why you are able to work on a computer right now."

Dave looked at his computer and then scratched his temple. He leaned slightly forward. He told me that for him, it was all a lot simpler. "I don't think that society is trying to coerce us," he said. "I don't see it like that. There are expectations we set for others that make life better for everyone."

I was skeptical of his direction. It sounded like he was going to be too diplomatic.

"Like we don't let people shit on buses," he said, "because no one wants to be around shit on a bus."

Did I say "diplomatic"?

"That's how I see sweat," he said.

"Like shit on a bus?"

"Yeah," he said. No one, he told me, wants to be sniffing other people's BO.

In that moment, I recognized that maybe I wouldn't ever be able to sweat and have it not be weird, because sweat *is* weird. The translucent liquid arrives on its own accord and has so many different meanings; it's not only a bodily function, but it can also inadvertently be a communication (or a miscommunication) about our emotional states. Sweat is gross, sexy, desired, loathed, and essential all at the same time.

It is also, apparently, quite a lot like shit on a bus.

Dr. Armpit

I was interested in learning more about the little creatures that eat our sweat and call our pits home sweet home, so I reached out to "Dr. Armpit," as a man by the name of Chris Callewaert is known in professional circles. Dr. Armpit deals almost exclusively with these axilla bacteria, and because it is these little guys that transform our innocent secretions into stank, he is therefore also on the cutting edge of nasty body smells. His career goal is quite noble, up there with inhabiting Mars, extending life, and creating renewable energy sources. In his own words: "I want to solve body odor."

Dr. Armpit has worked in this field for only a short time, but he has discovered quite a few fascinating factoids. Staphylococci and corynebacteria are most often the dominating strains that live in our pits. Staphylococci create a mild smell while corynebacteria can make the owner of a nearby unassuming nose yearn for a clothespin.

Those are the basics, but from there, Dr. Armpit found one of the four cornerstones of female superiority. Scientifically, women often stink less than men (though I don't think this is true of me—I can develop the rankness of a dirty sock left in a moldy basement for a week, but usually only in my right pit and after drinking a cup of coffee). This is because typically women's pits are staffed predominantly with staphylococci, while men, because of their thicker skin and fattier secretions, tend to attract the more gross corynebacteria, which thrive on a rich and oily diet. We can think of them as the sumo wrestlers, in appetite not necessarily appearance, of the armpit's bacterial kingdom.

Dr. Armpit has also conducted research about the effects of deodorants and antiperspirants. They don't cause Alzheimer's (as many still suspect), but surprisingly, in certain cases, they can make us smell worse. In a study, he found that consistent use of these products should not be a problem, but when used sporadically, they can actually increase the diversity of bacteria. That doesn't sound bad until you consider that one of the new species to move in might be the type that creates an awful stench. Also, antiperspirants tend to kill off more of the good-smelling bacteria, leaving the fetid guys the opportunity to take over. Dr. Armpit's takeaway: "If you don't have smelly pits," he warned, "I wouldn't recommend messing with them much."

In the future, Dr. Armpit hopes to make deodorants and antiperspirants that will specifically target the bad-smelling bacteria rather than meddle with the whole shebang. "We will be able to steer the microbiome," he said, "in a good direction."

Now this is where it gets really serious. Usually, Dr. Armpit explained to me, one's armpit microbiome is stable—the stalwart bacteria fend off any incoming invaders—but there are times when it can become unbalanced—he's heard of it happening in various cases such as when someone is ill, pregnant, on vacation, taking hormones, or staying in a hospital—at which point it becomes possible to inherit someone else's bacteria and therefore also their particular armpit funk. Yes, I'll say it again because I needed to hear it three times myself: There is actual proof that you can indeed catch someone else's BO.

The horror!

"It's not that easy," Dr. Armpit said, trying to assuage me, "but again, it can definitely happen."

This nightmarish prospect luckily has an upside. If bad BO can be transferred, then hypothetically, good BO can be as well. For people who reek so badly they are afraid to leave their homes (of which Dr. Armpit assured me there are plenty), he has begun experimenting with bacterial transplantation. It's like a heart transplant, only better because no one has to die and there's no blood. The donor refrains from washing her armpits for four days to ensure a plethora of fauna to harvest, while the recipient washes daily with an antibiotic. "Then we do the transplant," Dr. Armpit explained.

I got the visual, and quite enjoyed it, of Dr. Armpit standing in an operating room. He is dressed in a long white lab coat, face mask, and gloves. A lone armpit is below him on a gurney, and instead of being armed with a scalpel, he has one tiny cotton swab he applies meticulously, like a mom putting peroxide on her daughter's scraped-up elbow.

So far, Dr. Armpit has performed transplants on nineteen people who have all had varying levels of results. "We've noticed improvements," he told me, explaining that a few of his patients, to this day, continue to smell fresh. "But there is still a lot to learn."

While he continues to work on eradicating body odor—an effort that will surely win him the Nobel Peace Prize one day—he hopes that people refrain from becoming overly judgmental about their or anyone else's armpits. "If you have body odor, don't get too upset about it," he said. "Remember, it's not your fault, it's the fault of your bacteria."

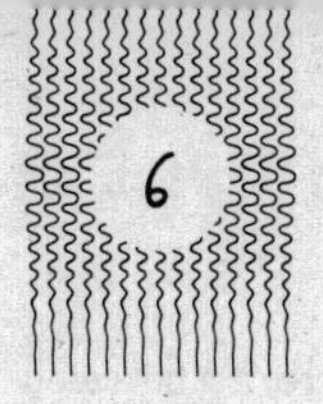

My Cup Runneth Under

On one recent humid September day, I was in my apartment bathroom, plucking my nipple hairs. During typical nipple-hair conversations, I tell friends that I have so many, like four strands. I have more like twenty-three really, but I round down. So I was standing there with my tweezers for about two minutes, but actually more like ten. I was there, sprucing up my breasts, because I was getting ready for my first urban topless bike ride. Hundreds of New York City women were gathering within the hour to exercise chest equality. In their minds, breasts are not sex objects that exist just for men to ogle. They account for our fifth and sixth extremities, which have been unjustly held captive inside rayon-blended bra cups for too damn long. They deserve, just like male nipples, to feel the direct caress of our city's balmy car-exhaust-laden air.

The theory goes that once enough people are regularly exposed to the female breast, it will lose its titillation. It will become desensitized. Normalized. On a scorching day, toplessness will be the new tank top. Breastfeeding will be a child snacking on some PB&J. Boobs, for better or worse, will effuse the intense eroticism of the elbow. Okay, they'll still probably be pretty hot, but they won't cause onlookers to get into a tizzy.

Breasts, as they stand (or sag or bounce), are still rather taboo out in public, so participating in something like this felt highly intimidating. I'd spent the two weeks prior tottering between "Hell, no" and "Probably not." First of all, I'm more or less flat-chested, and I suspected anyone who consciously chose to be topless in public would have a righteously large and perky rack. I'd feel like I'd showed up to a black-tie event in an awkward-fitting freebie convention shirt. Second, when I'd thought about participating, the possible pitfalls had seemed huge. What if I changed careers? Say someday I want to be a politician. I'm running on a ticket of universal health care and weekly pizza parties for all, but then, during opposition research, a picture of my bare tits bobbling above a bicycle is revealed. I am no longer fit to serve. Keeping fabric less than a millimeter thick between your body and the world somehow preserves your integrity and makes you honorable, respectable, and capable of deep thoughts. Taking that little swatch of material away makes you a hussy.

Instead of ascending to one of the highest offices in the nation, I end up grinding snow cones at roaming county fairs and have a fleeting moment of fame in a sideshow.

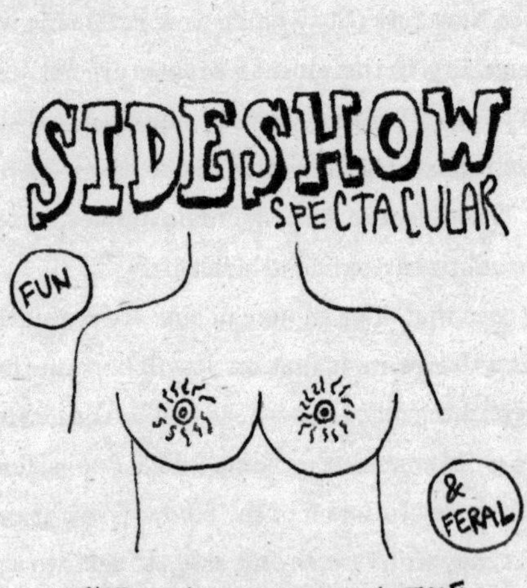

But maybe I was just being dramatic. We are now an open-minded and progressive society, right? No one would really care about bare breasts, especially in New York City. It's here that I've witnessed a man drop his pants on a subway platform and take a dump into an empty Starbucks cup and not one bystander even flinched.

So I asked Maggie, my very liberal and body-positive friend, if she'd join me on the ride.

"No," she said. "People will think I'm a sex perv."

She doesn't mind being a sex perv, but she understands that there are implications if others think she is a sex perv. Her perceived perviness could affect her business prospects as an upstanding acupuncturist. Clearly, you can't properly heal someone once they've witnessed your areolas.

After much deliberation—most of it done while picking three-month-old polish off my toenails—I decided to go ahead, throw caution to the underwire, and do the damn bike ride despite my reservations. I still wasn't sure about my own aims in participating. For one, I wasn't sure I wanted breasts to be normalized. What level would advertisers have to stoop to in order to sell beer and sports drinks? There'd be a whole cadre of horny teenagers wondering where to go after first base. Plastic surgeons would have to come up with a new profitable body part to inflate, and worst of all, I'd be left responsible to pluck my nips at much more regular intervals.

Still, I found myself too curious to pass up the opportunity.

As I stood in front of the mirror, giving myself a last once-over—jean shorts, knee-length socks, tennis shoes, bright pink fanny pack, and breasts with less volume than a minibar bottle of tequila—I wondered if the gathering was going to feel like the who-wore-it-best section in a women's magazine. All of us showing up to an event wearing the same (no) thing. Please vote: Did Mara Altman or Selena Gomez wear breasts best?

We want to be sisters on a mission, but are conditioned to judge—not only ourselves, but those by our sides.

Or would there be a sense of camaraderie, as if we'd all shown up in the same uniform like foot soldiers fighting for a shared cause?

After that revolutionary and empowering thought, I put on a shirt. Also, I put on a bra underneath that shirt. As absurd as it seems, I was concerned about exposing inappropriate nipple contours while on my way to a topless bike ride.

After a twenty-minute walk, I arrived at the meeting place, a bike tour company located just north of the Williamsburg Bridge in Brooklyn called Loudest Yeller. The business was in the brightly lit basement of a large brick apartment building. Only a smattering of girls had arrived, all with their tops still on and each suspiciously under forty. Did all the fortysomethings have more important things to attend to, or did wanting to expose oneself for equality expire after being hit with more than four decades of gravity?

About eighty people must have become lost on the way over, because the festivities were about to start and yet there were only ten of us. I was quickly informed that only ten had signed up. I'd planned on my own boobs being masked by masses of other boobs—one little fish getting lost among an enormous school—but we'd be more like the morning's catch displayed at a fish market. "Let's get ready, everyone," said Adam, a curly blond dude who owned the shop and who had volunteered to lead us around the city. "Pick out your helmets." He pointed to a wall where black and blue headgear hung from floor to ceiling.

The women began tugging off their shirts. You can never tell from the look of someone what their method of clothing removal will be—will they cross their arms, reach for the bottom hem, and lift it over their heads, or go for the much more precious style of taking one arm out of a sleeve at a time until they have a ring of shirt around their necks and then carefully hoist it away from their heads? I hail from the crossed-arm method and got ready to rip off my shirt, but before it rose above my sternum, it was as if I got caught in a glue trap. A pause button was pressed—my arms froze. I couldn't reveal myself.

Long before I had proper breast tissue or even Montgomery glands—those little white dots that make your nipples look like they're trying to say something in Braille—I knew that that area of my body belonged buried beneath layers of cloth. When I was twelve years old, my mom took me to get a bra at Nordstrom, never mind that I was still as flat as Kansas. I'd spent hours standing side-flected in the mirror, looking for any anomaly sticking up from my otherwise blazingly flat geography. I was a desert plane without the slightest topographical shift. No trees, no hills, nothing. I had a nipple, though—two, actually, which horrifically didn't deviate at all from my older brothers'. I didn't need a bra. Even though I wanted to need one, I knew that I wasn't worthy of one yet. Nevertheless, my mom dragged me into the lingerie department hell-bent on getting the help of a salesclerk.

This salesclerk, I was sure, was going to laugh at me. She'd stare at me, maybe pull my shirt collar out and look down: "I'd give you a training bra, but it doesn't look like you have anything to train!" She'd slap her knee and howl.

While my mom searched for this demon clerk, I ran off and hid in the middle of a circular clothes rack. Minutes later, the pink silk pajamas hanging in front of my face parted and there was the salesclerk pushing eight different tiny bras into my hands. For some reason, she found my desert plane absolutely acceptable. In other words, the training bra is not a device used to train breasts—to help lift them or support them—it is used to train us in covering ourselves up.

Meanwhile, Adam had already begun to fit the women to appropriate-size bikes. I had to get it together and be liberated already, so I turned around to face the wall. I crossed my arms—defying a lifetime of conditioning—and slipped off my shirt, a move that I have traditionally done only in a dark room after at least a second dinner date and two glasses of wine.

Then I turned back around and no one seemed to be looking at my boobs, but I felt like I was looking at everyone else's boobs. They must have been looking at mine, too, because, come on, they were boobs. Who doesn't look at boobs? Since it did not look like they were looking at mine, though, I hoped it didn't look like I was looking at theirs, either.

I had expected the most perfect boobs to show up. I would be the lone petite asymmetrical outcast, but it turned out that these ladies were all sizes and shapes—a nice full pair to my left, a droopy but perfectly lovely set to my right. There were areolas the circumference of baseballs and others a quarter would completely eclipse. There were ones like torpedoes and others like bags filled with sand. One set was located so high up that this girl, if there was a particularly bad pothole, would likely get choked out by her own cleavage. Bras, with their dome-like shells, make every pair appear so uniform. Bras are like wrapping paper, hiding a wholly surprising gift inside.

At this point, I had already been topless for about three minutes, and yet it was incredible: It still appeared as though everyone was still looking at me directly in the eyes.

Instead of being concerned about exposing themselves, the women had the probably much more legitimate concern about staying upright on the bikes while zooming through one of the most congested cities in the country.

"I don't even know how to ride," one of the women said.

Adam waved me over and placed a blue helmet down on my head. "Fit?" he asked.

"Yes," I said.

He pushed a yellow-frame bike toward me and then told me to bring it outside to give it a test ride.

The glittering sidewalks and concrete buildings effused the dense wet heat of a sauna. Modesty, I quickly came to realize, wasn't the only reason people wear shirts. One is for sun protection and the other is to absorb your (or, in this case, my own) profuse sweat.

Even though I was in a more natural state than I normally am, I felt like I had to try harder than normal to act natural. If I put my hands to my

I EXPECTED ONLY THIS:

BUT GOT SO MUCH MORE:

sides, I felt like I was trying too hard to appear like nothing was out of the ordinary. If I crossed my hands over my chest, I felt like I was trying to hide. Hands on the hips just seemed too predictable. I settled the conundrum by getting on my bike already.

By that time, all ten of us were now on our bikes, looping up and down the empty sidewalk, giving the brakes a test run. And we all still had boobs. I made sure to double-check.

Conundrum: Even though I have my own pair that I can stare at anytime I want, I still kept looking at everyone else's breasts.

Conclusions:

1. We have a long way to go before desensitization.
2. Boobs are inherently distracting.
3. I'm developmentally stunted.

I counted a pair or two or three that I wouldn't mind having as my own, but that's not creepy, right? It's just like admiring someone's long legs or toned biceps.

In fact, in that moment, being surrounded by breasts was doing less to normalize them than to make me think about them constantly. Maybe the desensitization process is like the flu—it gets worse before it gets better?

It was early yet, but so far, the only real benefit of going topless that I could gather was that getting dressed becomes 80 percent easier. Breasts don't clash with anything. When I surveyed the group, I saw jeans, red culottes, gray pedal pushers, and even black sailor shorts. Everything worked with nipples! I think we, as a society, would save so much time each day if we only had to pick out pants.

Adam, who in solidarity was also bare-chested, waved us over. We all began pedaling toward him. In honor of the topless ride, he told us that we'd be visiting various feminist locations throughout the city. "Let's do this," he said, dipping his wheel into the bike lane.

So far, we'd only been on a quiet street, so we hadn't encountered any pedestrians, but that wouldn't last for long.

We began rolling down the asphalt single-file. Cars whipped past, and as we rode under a long line of trees, the heat of the sun flashed on and off my skin. The breeze flowed over my body, cooling me where little beads of sweat had sprung. I hummed as I felt my lips stretch wide.

But the peace was short-lived.

Less than ten minutes in, a man rode his bike alongside me. I thought he was going to say, "Nice tits," or something, you know, offhand like "Wow, how is it that your breasts are so hairless?" and I was getting ready to say, "I appreciate the compliment—I plucked this morning actually, but this is about equality and I am not a sex object." But instead he snarled and said, "Put a shirt on!" Before I could react, he took off and continued giving his message to each girl in front of me.

I remember when I first realized, like really realized and processed, that women had breasts. I was eight years old. (Maybe it took me so long to register because my mom had tiny knockers.) I was with Nancy, a middle-aged woman who took care of me after school. We were filling up at a gas station. It was when she went to pay the attendant that I noticed it—she had breasts like bowling balls and walked like she was lugging around two suitcases on her chest. It was odd—these bouncing balls of flesh. These things that presumably one day I would have. I wondered how women ran. How they wore seat belts and jumped rope. Did these things tug and hurt or, worse yet, become tangled? Ultimately, I wondered how women lived life with these masses—didn't they get in the way?

As I watched the man pedal off ahead of us, it became clear that I was onto something as a child, but I'd had it slightly off. Breasts don't get in the way in the logistical or tangible sense that I'd expected, but sometimes in a much more insidious way.

At about twenty minutes in, we crossed the Williamsburg Bridge into Manhattan. I was near the front of the pack, and as we neared the base of the bridge, there were three people with camera lenses that looked like an extra limb sprouting from their foreheads. I'd seen the same kind used by photographers on the National Geographic channel when attempting to capture polar bears frolicking on snowy outcroppings. They'd been tipped off. These three people had press credentials around their necks and were aiming straight at us.

When toplessness is newsworthy, it's safe to say it's a long way from normalization.

There went the presidency.

We rode for two hours before lunch. We rode through Battery Park and stopped in view of the Statue of Liberty, where one of the girls—one of the ones who had breasts that I wouldn't mind having (it was something about the perkiness, the lightheartedness of the pair, like they were tulips reaching for the light in the sky)—read Emma Lazarus's sonnet "The New Colossus."

HER: A mighty woman with a torch, whose flame is the imprisoned lightning, and her name Mother of Exiles.

ME: *I wonder how far off my breasts are from looking like her breasts.*

HER: From her beacon-hand glows world-wide welcome; her mild eyes command the air-bridged harbor that twin cities frame.

ME: *If I had breasts like that, I wonder if life would be easier.*

HER: "Keep, ancient lands, your storied pomp!" cries she with silent lips.

ME: *If I had breasts like that, I think I'd want to be in the newspaper.*

After the recitation, we made a quick stop at the New York Stock Exchange, which was filled with men in suits, but what really stood out were the many tourists aiming their cameras at us. Our meaningful movement, to them, was merely a stunt to document on their Instagram feed.

We then stopped at a building near City Hall Park where Susan B. Anthony and Elizabeth Cady Stanton had started a women's rights newspaper called *The Revolution*. I wondered whether, if Susan were alive today, she'd be down for Free the Nipple or think it frivolous while women's genitals are routinely mutilated in countries like Guinea.

There were more stops, but what really stood out—more than being half nude in traffic or the rare sensation of exhaust hitting my bare chest—were the reactions of the people we passed. The ride seemed less an expression of our rights than a roaming focus group meant to collect bystanders' thoughts about bare-breasted women.

People shouted, "What is this about?" and some said, "This is terrific," while others yelled out, "Hey! That's illegal."

Going topless as a female is not illegal. I didn't know it before finding the Outdoor Co-ed Topless Pulp Fiction Appreciation Society, the organizers of this ride. Turns out I could have been sipping lemonade in Central Park with my nips out since 1992, the year female toplessness became officially condoned in New York.

I saw interest, shock, disdain, adoration, and curiosity on the faces that flashed past. Many, hoards in fact, turned their phones toward us and began recording. I got it; usually, viewing this kind of stuff costs money and endless viruses on one's computer. I tried to be chill. I pretended, like most celebrities do in our nation, that since I had my hat and sunglasses on, my face wouldn't be recognizable. But every time a camera pointed in our direction, one of the girls, the one who had grandiose breasts, large and pillowlike, the type I'd decided would be perfect to rest my face in for a quick respite from the world, would yell, "Fuck you, you have to ask!" while flipping the person (or depending on the case, the crowd) in question the bird.

I didn't share her sentiment. Going outside topless would be like going out with a pair of parrots chanting "I like big butts and I cannot lie" while fornicating on your shoulder, and expecting witnesses not to snap a picture. It wasn't realistic.

When we rode along the southern part of the island, a mother shrieked with horror and then spun her young daughter around to save her from seeing our bare and horrifying flesh. Two years earlier, that same young girl was probably suckling on a breast. In a few more years, she will have her own pair.

While there were many who opposed exposed boobs, there were also people who just as adamantly supported our venture. While we were stopped, looking at the site where the Triangle shirtwaist factory fire occurred, two older women with white curly hair and oversize visors peered at us from afar. The way they had their heads tilted and eyes squinted, I thought they were disapproving, but they soon came up to us and said, "You all are just so beautiful."

It was very sweet. They genuinely seemed to be referring to the asymmetricals as much as the symmetricals.

Also, tattooed people with dyed hair and piercings in painful places hooted and hollered their approval. Couples, both sporting thick-framed hipster glasses, did as well. Some raised their fists in solidarity. I felt so cool to have clearly very hip people support my decision to show them my breasts. I know the ride wasn't about being cool and getting approbation, but at that moment, I have to admit it: I was kind of cooler and edgier than all of them and it felt awesome.

But throughout it all, there was confusion as to how we should react to the crowds. As we rode through Little Italy, a guy with a beard and a tight white T-shirt turned and held up his hand for a high five. When I put up my left hand, my bike wobbled, but as I passed, we made contact and a wonderful clapping sound. It was so amazing—people train for decades to become lauded as professional athletes or skilled performers, but as it turns out, you can get all sorts of adulation by just taking off your shirt. I was fully soaking up the praise. In fact, I was wondering if by doing this ride I'd tapped into some previously unrealized potential to be a top-notch exhibitionist. Maybe I was going to turn into a boob-showing junkie. Only, I couldn't tell if it was the freedom of my nipples I loved or the adrenaline rush created by doing something so divisive.

In the middle of my reverie, one girl, who had an obnoxiously fantastic mammary situation, said, "You shouldn't have done that."

I thought maybe she'd finally discovered that I'd been looking at her breasts all this time. "Done what?"

"High-fived," she said.

"Why?" I asked. I had considered the move warm and wonderfully encouraging.

"Because we don't want this to be considered good or bad. We just want it to be normal."

She explained that we have won our cause once there are no reactions at all, once we can just ride down the street in our bare-bosom glory without shouts of approval or boos of disgust, the same way a man can jog in a park without a shirt on.

For being such an open-minded revolutionary, she was being so

I COULDN'T REMEMBER ANYONE'S FACE

judgmental. Before she shut me down, I would have been happy to Freaky Friday with her boobs, but now the thought of having her rack became a lot less enticing, especially if it came with that attitude.

Just kidding. I'd totally swap! I'd get so many high fives.

Next, we parked our bikes under a tree near Washington Square Park. We dismounted and took off our helmets—our faces flushed from the heat and our hair wet and clumped from sweat. Summary: We were disgusting and hungry.

We walked into Murray's Cheese, a deli and sandwich place. Everyone dispersed to different parts of the shop. I was off on my own, looking at a laminated menu, when a tall middle-aged man wearing an apron stained with a variety of condiments came near. "We can't have you in here," he boomed. His voice really did boom. It sounded like a thunderous warning before lightning was about to strike me down. "This is a health code violation."

My face flushed another layer of hot and my palms got the same sticky as day-old spilled beer. It was one thing to have people condemn you from afar in the street, but another to be sedentary inside a sandwich shop and have every person's head turn toward you. I could tell that these people were looking at my boobs, which made me worry that the others could tell when I'd stared at theirs, but never mind that right now; this was an emergency. I lowered my eyes and saw what some of the customers were holding—a baguette, an Orangina, a block of cheese. It was suddenly quiet; I heard a squeeze-bottle of mustard fart out a yellow line.

I tried to nonchalantly cross my arms over my chest as I began scanning the store to see what the other girls were doing, but not one of my bike-riding friends was around anymore. Without anyone there, my reality shifted. What does a uniform mean when you're the only one left wearing it?

I dropped my backpack to the ground, unzipped it, and plunged my hands inside. I stirred the contents around, hoping to find the soft cottony texture of a T-shirt. I pulled it over my head desperately, like I was trying to pull myself out of a riptide. With the shirt on, I could finally breathe again. I went straight to a basket of ready-made sandwiches, turning them over to see the ingredients, but was actually blind to the labels as I attempted to make myself look busy and unashamed.

I learned later that upon getting kicked out, the other girls had left immediately, in a huff, and gone a door down to a topless-friendly deli. That all happened somehow while I'd stood frozen—while I went directly from confident exhibitionist to cowering shame-bot. Did it really take the condemnation of only one dude in a mayo-splotched smock to spur such a drastic transformation?

I met the others out in the park on a slightly damp section under a large tree. They were all still lighthearted and laughing; one was even playing kickball with some young kids. A couple of girls were lying in the grass while others were taking big bites of their subs. Two scooted backward in order to widen the circle, giving me space to sit down. I unwrapped my sandwich and then took off my shirt again, this time with much more ease and determination than I had a few hours before.

After lunch, we rode for two more hours, much of the time through the

West and East Villages, before hanging up our helmets. The same hooting and hollering occurred—grimaces and glares mixed with thumbs-ups and attempted high fives. We made a few more stops before a late-afternoon rain broke the thick heat, and as little droplets pinged pleasingly all over my body, I finally realized an interesting change—my breasts, in that moment, weren't for anyone but me.

I hadn't really dwelled on it before, but since my beginning, my breasts have always been for someone else. When I was a teenager, I wanted my breasts to grow so I'd be attractive to boys. When my breasts turned out small, I felt it my duty to warn boys before they went under my half-filled bra cups so they wouldn't be disappointed by what they found. For doctors, my breasts were something that could potentially turn lethal. For the babies I may have one day, they would be a source of food. For fashion, breasts are a way to accessorize. Being topless is always a stop on the way to somewhere else—to a shower, to a breast exam, to sex—but is rarely the destination in and of itself. By exposing my breasts to everything and everyone in one of the largest cities in this nation, paradoxically I finally got a taste of what it was like to relish them for myself.

The next day, I woke up and immediately flipped open my computer. My heart beat hard as I looked through all the newspapers. A week before, I would have expected my accelerated pulse to have come from the fear of being exposed, but today it stemmed from excitement and pride. In the *New York Post,* I found the photo. It was on the first page of the local section under the headline "Breezy Riders." I would have gone with something more sophisticated like "Boobies Everywhere!!!" but hey, it wasn't my operation. I was near the front of a group of four of us with my blue helmet and knee-high socks. My breasts were censored with black bars.

They can show the war dead, but not the female nipple.

My black bars were the same size as the other girls' black bars, though, which I found gratifying. Equality, at last.

Actual Navel Gazing

I was recently at a coffee shop with my mom. While we waited for our coffees to cool down, I finally asked her something I had never dared to bring up before. "How did you feel when you first saw my belly button?" I was talking about the moment my umbilical scab came off and she could see what lay beneath.

"I just kept repeating to myself, 'It's okay, she's going to grow into it. It's okay, she's going to grow into it.'" Then my mom got a weird smirk on her face and burst out laughing. "It was just so big," she said, demonstrating by holding up both her hands. She was being dramatic—a kitten could have fit in the empty space between her palms. "You had an adult-man belly button on your baby body." By this time, she was laughing so hard that she was wiping her eyes with the hem of her long-sleeved shirt. She could hardly speak. I wasn't sure why she was so hysterical. Maybe this was what catharsis looked like after decades of keeping a straight face. "Your brothers got these great dainty little divots," she said. "I was like, 'Why did my daughter have to get the giant protruding one?'"

"But when I was little, you always told me that it was okay," I said.

Her eyes grew big. "Well," she said, "what was I supposed to say?"

She's one of those people you may know: as she's gotten older, she's completely lost her ability to politely lie, even for the well-being of her loved ones. After taking a sip of coffee, she suddenly tried to get all mom-like and supportive again. She rubbed me on the shoulder. "It turned out okay, didn't it?" she said.

When I was a kid, my shirt stuck out from my stomach like a tent—my navel was the pole. It was a permanently erect abdominal nipple. It was irritating; it rubbed on everything. I couldn't even do Slip'n Slide properly; the button hurt and caused drag. Friends, during swim parties, would point at it and say, "What's that?" as if I may have been harboring a miniature Kuato inside my bathing suit.

Every other week, I'd ask my mom if she could chop it off. It was usually at convenient moments, like when she was already dicing onions on a cutting board.

"No, not until you're eighteen," she'd say.

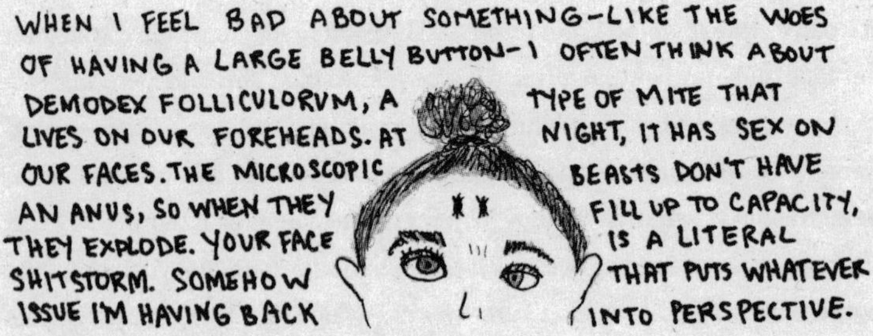

WHEN I FEEL BAD ABOUT SOMETHING—LIKE THE WOES OF HAVING A LARGE BELLY BUTTON—I OFTEN THINK ABOUT DEMODEX FOLLICULORUM, A TYPE OF MITE THAT LIVES ON OUR FOREHEADS. AT NIGHT, IT HAS SEX ON OUR FACES. THE MICROSCOPIC BEASTS DON'T HAVE AN ANUS, SO WHEN THEY FILL UP TO CAPACITY, THEY EXPLODE. YOUR FACE IS A LITERAL SHITSTORM. SOMEHOW THAT PUTS WHATEVER ISSUE I'M HAVING BACK INTO PERSPECTIVE.

As I got older, what my mom suspected—that I would grow into it—was true, though instead of growing into it, I actually grew around it, my belly ballooning on all sides of the nub like muffin batter rising around a chocolate chip. Now, when viewed straight on, my stomach appears to be birthing a tiny pink baby mouse.

My sister-in-law, during a moment of creative inspiration, said it looks like a chewed-up piece of bubble gum.

Besides having an outie, my knob is also inordinately sensitive. When I was growing up, my brothers knew that touching it was an easily accessible form of torture. Press it and it detonates a missile of nausea. I had to teach boyfriends that it was a no-go zone. I'd say something understated like "If you touch my belly button, I will rip off your eyelids and use them as toilet paper." Anything subtler than that seemed to be taken as a coy invitation to attempt to make contact.

As an adult, I'm able to display my navel without embarrassment, and I've even had a friend touch it without my vomiting on her face. But because of the shape of my navel and the unsavory sensations it inspires, I've given the whole situation an unseemly amount of thought. I've concluded that the navel is obnoxious; it's prominent, but it doesn't even do anything. The navel is for . . . it is for nothing.

After working double duty for the first nine months of our development, as both our mouth and anal orifices—two highly respectable roles—at the moment of birth, it gets a gigantic demotion from CEO of survival to weird stomach indent. By the time we become conscious of it and start asking questions, the body part is already retired. Parents try to tell you it's a button, but nothing happens when you press it. Then they tell you that you used to eat through it, but when you hold up food, it doesn't take a bite. Our other useless parts sometimes surprise us with pain—I'm looking at you, wisdom teeth—but at least when they do, we dispose of them. The navel just stays there forever, right there in the middle of our stomachs. Not only don't we remove it, but sometimes we even turn it into a fashion statement.

Maybe it's the navel's pointless yet permanent presence that leads scientists, laymen, and even entire cultures to desperately try to make sense—justify the existence—of this derelict hole.

Desmond Morris, famed zoologist and author of *The Naked Ape*, believes that we retained our prominent stomach crevasse while other mammals, such as dogs, leave virtually no umbilical evidence behind, because for us upright-walking humans, it's a useful "genital echo," a kind of symbolic "pseudo vagina." Theoretically, the navel helps remind the opposite sex of our other more reproductive organ a tad bit south. I guess that might work, except if yours, like mine, looks like a little otter penis.

In Chinese acupuncture, the belly button is referred to as the Spirit Palace. Puncturing it with a needle is prohibited, but if you put salt and ginger inside, it is supposed to cure diarrhea. In some Eastern traditions, the navel is also considered the location of an important chakra. Guru Mantra Singh, a kundalini yoga master I contacted, told me that the chakra embodies vitality, energy, and power.

To explain how it works, he asked if I had a puppy. "If it pees on your rug," he said, "you speak to it from your navel." Then he demonstrated. "No, don't pee on my rug!" he yelled. "Get it?" he asked.

Not really.

There are plenty more navel interpretations—a *Time* magazine article from 1959 tells me that in early Japan, if a baby was born with a navel pointing downward, "the parents would brace themselves for a weakling child who would bring them woe." I even spoke with a man, Jonathan Royle, who uses the navel to tell one's fortune. When I sent him a picture of mine, he wrote back to tell me that people who have buttons like mine "tend to have split personalities and are often very unpredictable and often emotionally unbalanced indeed."

I wonder if that's what my husband is referring to when he tells me it's not normal to give motivational talks to my undersunned houseplants.

All these perspectives were fascinating—if not a little desperate to make meaning—but for me, the navel started to make sense only once I spoke to an anatomist. The tangible and concrete facts added up to a truth that I couldn't have seen otherwise.

Derek Harmon spends most of his time wrist-deep in cadavers at the University of California at San Francisco, where he teaches human anatomy to medical school students. Before digging too deep, I wanted to find out the truth behind an outie. How an outie is formed—as benign an issue as it seems—is oddly a hotbed of conspiracy. Harmon, a professional, even had to powwow with his co-anatomists to discuss the reason behind our differing shapes.

Before I spoke with him, I had been exposed to a lot of hearsay. A friend thought that the ob-gyn takes a look at the mom's navel and then snips the baby's so that they'll match. Though it's a popular theory, I already knew that an ob-gyn cannot take credit for shaping the navel when she cuts the cord any more than a mohel can take credit for the size of a penis after clipping its foreskin. I knew that because someone once pointed at my abdomen and said, "Don't you wish you could go back in time and tell your mom's doctor to pay more attention?"

I'd approached my aunt Karen, an ob-gyn and a professional cord cutter, about it, and she had told me that navel shape wasn't the fault of her kind. Besides, many mammal mothers chew off the umbilical cord of their young and don't do such an immaculate precise job, yet you don't see big honking stalactite button structures hanging off a cow. The umbilical cord is cut, and the remainder—whether it's a centimeter or five inches—will shrivel up into a scab and fall off at the abdomen.

My mom, when I'd asked her on several million occasions why I got the one I did, always said, "Your dad has one, too."

It's true; he does. Because of that, I'd always believed my navel size was due to heredity. I had a genetic outie.

Harmon told me that I was wrong, too.

To describe what happens, he used balloon analogies and words like *gradient* and began with a very unrelatable statement. "It's related to the presence of space between the skin and abdominal wall when you're born," he explained.

After a much longer conversation, I began to understand. In a fetus, a wad of skin forms at the mouth of the navel. Now, think of your abdomen as a balloon and the navel opening as the part of a balloon you would tie off to keep all the air inside. Most of the time, that knot stays tucked inside. "An outie happens because there is so much pressure in the abdominal cavity that it pushes the skin outward to where it kind of pops out of that top part of the balloon," Harmon said.

So after all this time, I found that the difference between me and an innie was a mere air pocket. No more than a fart bubbling up in a pool. A bit of carbonation. It was underwhelming. Then again, there was something equalizing about this information: All of us are outies—innies are just outies that haven't been ejected yet.

My whole life, I suspected that the navel was purely aesthetic, a standalone nodule existing only on the surface, but it actually has deeper roots. "Unlike what most people would assume," Harmon said, "the underside of the navel is not a flat surface."

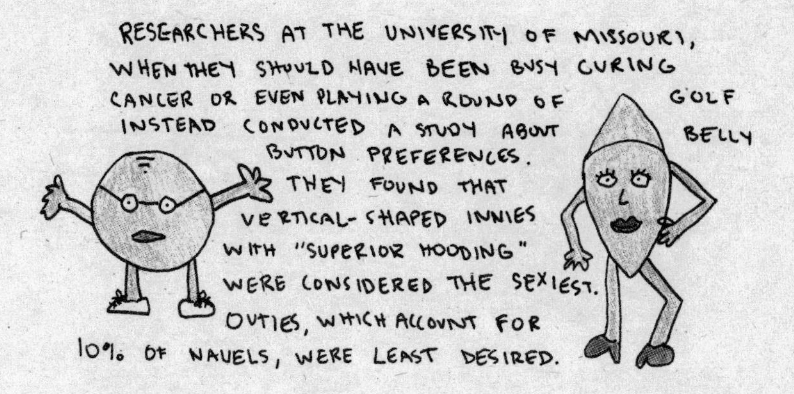

When the umbilical cord passes through the navel and into the abdomen, it fans out from the cordlike structure into a network of arteries, which, while in utero, keep us pumped with blood and nutrients. When the cord is cut, all those underlying structures remain. They dry up, close off, and turn into ligaments, but they remain attached to organs such as the liver and bladder. These ligaments are like condemned buildings—they still stand but no longer have purpose.

In fact, your belly button, as you read this, is attached to your bladder via a defunct structure called the urachus. "When I open up the abdominal wall of a cadaver," Harmon explained, "it looks like four to six hairs of pasta are coming out of the back of the belly button."

Our navels are only the tip of an underground weblike iceberg. And guess who's not going to eat spaghetti this month? Me!

In extremely rare cases, these abandoned ligament structures don't seal up after birth. If that happens, a condition may occur where you can pee through your belly button. Can you imagine the convenience (and the adorable tiny urinals)? Unfortunately, this is not something that you can learn to do on your own. Doctors actually consider the ability to pee through your belly button a problem.

There is another surprising thing the button can do: A sudden onset of belly button pain might signal that you have appendicitis. This is because

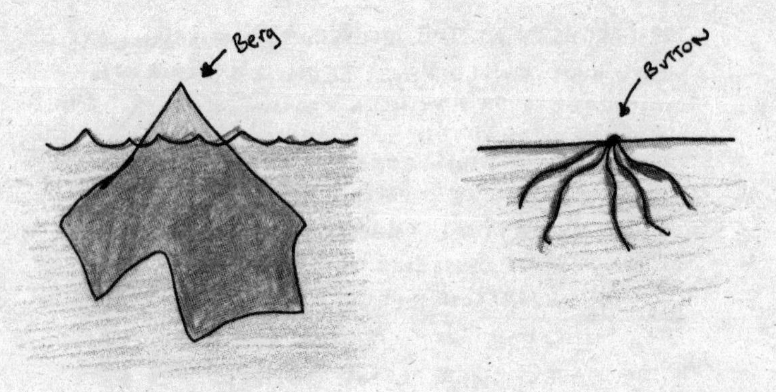

nerves from the navel share the same part of the spinal cord as the sensory neurons that supply the appendix.

In other words, when the appendix hurts, the brain gets confused and first sends pain to your navel. This is the lesser-known ugly duckling stepsister to the more popular "referred pain" experience that occurs just before a heart attack—people first feel a pain in their left arm. It may be a tad dramatic, but I think it's defensible to say that a belly button can actually save your life.

I was feeling pretty satisfied with this onslaught of information, but Harmon even had a scientific answer to what I thought was my mysterious and possibly supernatural lifelong queasy sensation.

"It's such a sensitive area," he began.

Though the belly button, as discussed earlier, suffers a gargantuan demotion after birth, the umbilical cord was once our lifeline—if anything happened to disrupt the connection, our fetal lives would be in peril. Because of this, the area is saturated with abundant nerves—rigged like a supersensitive alarm system. "If the nerves were tripped," said Harmon, "they'd warn your brain, 'Hey, something is going wrong. Watch it!'"

Though there has been no formal study, Harmon suspects that as we grow, the nerve connections can make for unpredictable sensations—hence the reason I can feel nausea while someone else, from the same navel touch, can feel pain, ticklish, turned on, or even nothing at all.

"I hate my belly button getting touched, too," Harmon admitted.

My queasiness then was not a fluke. The feeling was because of an anachronistic warning system, like a land mine still buried long after a war, which was blindly doing its best to help win the battle for life. My body—behind by more than three decades—still didn't want anyone screwing around with my no-longer-existing cord.

While the navel can mean many different things to different people—a spirit gateway, a chakra, a symbol of fertility—it is also undoubtedly, Harmon told me, and across the board, our first scar. Whether we have an overpressurized outie or a vaginal echo of an innie, when we are born, the area begins as a wound that must heal. It's not just a retired body part or a handy landmark used to identify the abdomen; the belly button is actually our first flaw. It's our body's way of telling us we're only human.

On Belly Button Lint

I don't produce belly button lint. Nothing. Never. I am therefore quite jealous of people whose buttons do collect lint—soft little navel pearls, slowly assembled from the finest grains of weird and random shit.

I asked my husband how he feels when he finds that his button is fully loaded.

"It feels like an accomplishment," he said. "I didn't necessarily plant the seeds, but I get to harvest."

The few times that he let me pluck the fuzz, I felt like he won me a stuffed animal at the county fair, only it was much better because I could throw away the lint instead of pretend that I was excited to have a gigantic stuffed anthropomorphic banana take up the majority of my bed for the next two years.

That being said, I did not think there was much mystery behind belly button lint: There's a hole and crap gets stuck in holes. No different from how trash collects in highway potholes.

But there happens to be researchers who weren't satisfied without knowing the specifics (e.g., why do some potholes collect more trash than others?), so they decided to study exactly how atmospheric dreck gets lodged inside the belly's black hole.

Spoiler: Lint is not, as you might have hoped, a tiny blanket knitted by your navel microbes.

Dr. Karl Kruszelnicki, who hosts a science show in Australia, launched an informal study in which there were 4,799 people who responded to various questions about the status of their lint. When Dr. Karl calculated all the answers, he found that the most prolific lint producers tend to be hairy overweight middle-aged men with innies. In other words, if we were to start a belly button lint factory, we would want to do most of our recruitment at all-you-can-eat buffets and the Rotary Club.

Dr. Karl found that body hair, most likely, acts as a vast network of railways that channels debris toward the navel depot. Some lint may even be traveling from as far as the underwear region, where it catches a nonstop ride upward on the "happy trail."

Meanwhile, all the way across the world in Vienna, Georg Steinhauser was busy studying his own lint production—collecting and weighing each piece—for three painstaking years. He came to the same conclusion and published his navel-fluff findings in the journal *Medical Hypotheses*. "The scaly structure of hair firstly enhances the abrasion of minuscule fibers from the shirt and secondly directs the lint into one direction—the navel—where it accumulates," he wrote.

To test his hypothesis, Steinhauser shaved his stomach and thereby destroyed the lint's vast transportation network. He was, in fact, onto something: Until his hair grew back, not one piece of lint came by for a visit.

Steinhauser went one step further (some might say too far) and analyzed the chemical makeup of his lint. He found that it was made up not only of cotton fibers but also of house dust, cutaneous scales, fat, proteins, and sweat. He surmised, perhaps a bit self-servingly, that lint producers (i.e., hairy fat men) have more hygienic belly buttons. "Lint helps sweep it clean," he concluded.

So a navel full of lint, it turns out, is not gross; it's actually in self-cleaning mode.

The
Bottom
Half

The Air Down There

When I brought home my new puppy a year ago, I quickly discovered that she was drawn to one thing more than all the other things. She was a crotch junkie. From every corner of the house, she'd be called forth spontaneously to come and sniff my vulva.

Chucho is not the only dog that has been fascinated by my crotch. I have been confronted by quite a few crotch-sniffers in my life. The time I recall most clearly was when I went to interview a woman for an article and her dog spent the first five minutes of our sit-down with its head between my legs. It's hard to appear sophisticated and professional when a large shepherd mix is trying to inhale your ovaries through your jeans. I wanted to point to my crotch and tell my interviewee, "I just want you to know, there is nothing odd or overtly pungent about this crotch," but I wasn't sure that was true. I had no idea what made a dog go for one crotch over another, so instead I tried to pretend the whole thing wasn't happening as I shoved my notebook down there, turning the block of paper into a makeshift chastity belt.

No matter how awkward those encounters have been, it had been easy to ignore the phenomenon because it happened only sporadically and always outside my home. I could get in my car, drive away, and be in denial about it ever having occurred. I didn't have to ask what it meant about me or ponder the greater subject of vulvicular (not a word, but it should be) aromas and how they fit into the smell-scape of modern urban life.

When I got Chucho, though, I could no longer ignore the situation. A dog snout was near my groin so often that it seemed like I'd acquired a new genital attachment. I couldn't even safely sit on the toilet anymore. Three weeks into her tenure at our apartment, she charged into the bathroom when I was at my most vulnerable and locked her jaws onto the crotch portion of my underwear. She dug her front paws into the bath mat as she tried her hardest to pull my underwear free from my ankles. I tried to shake her loose, but only so much movement is advised in that situation. Plus, I'm bad at multitasking.

I yelled to Dave for help. Even when you're married and you've shared just about everything, you don't necessarily want him to see you on the toilet, but I was out of options. When Dave saw me in my predicament, instead of helping, he started laughing so hard that he was paralyzed. I haven't seen him that enthralled since *The West Wing* became available on demand.

"Help me!" I yelled.

He couldn't get it together. That dickwad was enjoying the struggle.

The whole fiasco wasn't worth the teachable moment I'd wanted it to be for my puppy, so I finally gave up and let her have my underwear so that I could finish my dump in peace.

She dashed—underwear between her teeth—toward her green zone, which was the unreachable back corner beneath our bed.

I could have taken my pup's crotch addiction as a compliment, but as I've watched her grow, I've noticed that her other favorite scents include feces, vomit, rotting garbage, and dead rodents. I feel like she is passive-aggressively communicating to me: "Your vagina smells divine . . . just like decomposing possums."

Being constantly reminded that between my legs there was an odor, something potent enough to attract furry animals, brought up both the insecurity and the fascination I've had about vaginal scents over the years.

When I take a whiff of myself, I think it smells, well, vagina-y. Like it's supposed to, I guess. I actually think it's rather impressive, considering it's an open orifice and only inches away from an active butthole. That being said, it *is* an open orifice and only inches away from an active butthole! I wanted to know the role, if any, vag scent played in my life.

I jumped off on this exploration exactly where my interest first became piqued: What was it about the human vulva that attracted dogs?

Believe it or not, it wasn't easy to find any trained scientists who were open to sharing their knowledge about why a dog might like to huff vaginas, but after a few false leads—"No, I have not conducted research about human crotches as an attractant to canines"—I finally found one.

George Preti is a chemist at the Monell Chemical Senses Center, and in his long and illustrious career, not only has he studied armpit secretions, but he has also trained dogs to sniff out ovulating cows to help ranchers more successfully inseminate the heifers. Currently, he's training dogs to detect cancer.

He graciously took time away from saving people's lives to let me ask him some questions. "Why does my dog go gonzo about my crotch?" I asked. There was something embarrassing about this occurrence. I didn't know if it was because my crotch was more overtly pungent than other crotches or if my crotch was like a sunset to her—she finds it so beautiful that each day she can appreciate it anew.

I was partial to the latter interpretation and was hoping that Preti would validate my theory. I'm reluctant to admit it, but it was also hard not to wonder if Chucho was exploring pansexualism and had an interest in cross-species relationships.

"It's the most odiferous body part available to dogs," Preti said. "If you were sitting down, they might be sniffing your armpits or your ear."

While his answer wasn't exactly flattering, at least he didn't say because your crotch smells like garbage and dogs love garbage.

He explained that those three areas of our bodies produce our individual odor profiles. "They communicate who you are, which sex you are, and probably even where you're at in your cycle."

In essence, every time my dog sniffs me, in her head she's thinking, "That's Mara." Ten minutes later: "That's Mara." Thirty-five minutes later: "Yep, that's still Mara." In that sense, my dog is like a kid looking across the playground every once in a while to see if her mom is still there, only except for innocently looking, she has to smash her snout against the most private and sensitive part of my body.

Then Preti said something awful. "There's urine there, too. There's always a bit of urine, and urine is very informative for a dog."

All I could think about was how I was probably going to cut this part out of the interview because I pride myself as a champion wiper and therefore there is obviously no way that urine could be part of the equation.

"So are they drawn to the strongest-smelling vagina or the one that is most intriguing to them?"

"I'm not a dog behaviorist," he said. "I don't know what they are thinking."

"But you're saying it's typical—that women shouldn't be embarrassed when a dog comes in for a whiff, right?"

"I'm a chemist," he said, "not a psychologist."

Preti was not being nearly as Liberal Arts Major as I wanted him to be, but what I gathered from our conversation was that we could certainly interpret a dog's interest in our crotches as a compliment. The dog wants to get to know us better. It's like a person who asks, "How are you?" but who won't just take a perfunctory "Fine" for an answer. They *really* want to know how you are—so much that they need to sniff your vulva.

If what Preti said was true—that the scent of my vagina, in an olfactory sense, is who I am—then I wanted to know what the composition was behind my particular aroma.

In other words, what exactly is responsible for making me smell like me?

To find out, I contacted Tiffanie Nelson, a vaginal secretion expert and crotch genius. She is currently a research fellow at the Geelong Centre for Emerging Infectious Diseases and has spent many years investigating vaginal tract microbiota. "What makes us smell?" I asked Nelson. She deals exclusively with vaginas, so I didn't even have to distinguish which body part.

"Hold up a moment," she said. She explained that we had to start a bit further back. She said that our vaginal scent is closely linked with the billions of animals that couch surf down there. A healthy vagina is teeming with bacteria, the majority of which are lactobacilli—the same kind of bugs we use to make sauerkraut and yogurt. "They perform the same function as they do when we ferment foods," she said, "but instead they do it in vaginas."

She explained that our vaginas are like an all-you-can-eat buffet for our bacteria. Our vaginal skin constantly excretes sugars—think Slurpee machine stuck in the on position. The bacteria chow down, turning the Slurpee into lactic acid.

"So, wait—are the bacteria pooping inside of our vaginas?" I asked.

"They don't have buttholes."

"Then how does the lactic acid come out?"

"The sugars come through one side of the cell membrane, and then the lactic acid comes out the other side," she said. "I'd call it a 'flow-through.'"

"That sounds like a euphemism for pooping."

At that moment, it seemed crucial to know whether or not my genitalia moonlit as a toilet—that would be so meta.

"If you really want," she said, "you can call it 'pooping,' but that's not what it is."

I suddenly felt transparent about my psychological obsessions. "I might need to," I said.

She went on to explain that while we supply the lactobacilli with food and a place to live, we depend on their lactic acid to protect us—the acidic environment shields our crotch from the colonization of undesirable bacteria. Evil bacteria, entering a properly acidic environment, would meet a similar end as an astronaut who went to the moon without a spacesuit.

"So what does that have to do with scent?"

Though Nelson isn't interested in normal vaginal scent—she studies vaginas that are out of whack and smell like a pod of dying river dolphins—she said that when our vaginas are balanced, meaning a pH of 3.5 to 4.5 (within the range of apples and dill pickles), we have an "everyday vaginal odor." She couldn't go into detail about this scent because it doesn't even register on her smell-O-meter.

To find out what "everyday vaginal odor" meant, I interviewed many women, but most had trouble describing their scent to me. At least four women paused, looked up toward the ceiling, and then said something along the lines of, "I don't have the words." There were a few others who ventured a description: "Metallic." "Like a fine wine." "Musky and earthy." "I'd say briny, but not, like, overtly briny." "Earthy and yeasty." "Astringent." My favorite adjective was "Youwantmetodescribewhat?"

The odor is different for everyone and changes throughout our cycles, but it's the stuff that those douching companies would have us believe leaves us with that "not-so-fresh feeling."

I talked to a gynecologist, Dr. Jenny Hackforth-Jones, and she said that no matter the smell, if it's a healthy vagina, you shouldn't be able to catch a whiff of it if you're more than a foot away.

I asked if that rule held true even if, hypothetically, you biked for twenty minutes, took a yoga class, and then sat in your yoga pants for five hours while researching vaginal odors, and then got a beer at a bar and just happened to cross and uncross your legs a couple of times.

"A slight odor, right?" Dr. Jenny asked, concerned. "Not excessive?"

"Just vagina smell," I said. "What I think of as vagina."

"Someone couldn't smell it across the room, right?"

"As far as I know, I was the only person who could smell me."

"It's nice that you're doing yoga," she said, treading carefully, "but let things breathe, if you can."

I actually think that yoga pants—in all their overpriced and camel-toed glory—are out to sabotage us, but I'll save that for my book about conspiracy theories.

While we each have a natural scent, sometimes the vagina can get funky in a bad way. This usually occurs because there is a bacterial imbalance—outsider bacteria battle and begin to gain a foothold against the predominant lactobacilli—or an infection. The everyday yeast infection can make a vagina smell like it's whipping up a loaf of sourdough, which is why, when my friend's vag gets yeasty, her boy friend lovingly calls it "The Bread Factory." The much more offensive and notorious fishy smell comes from a condition called bacterial vaginosis (BV), a diagnosis that sounds like a code-blue genital emergency but is really only an imbalance of bacteria in the vagina.

In her most erudite scientific jargon, Nelson, my vaginal odor expert, explained the resulting scent: "It's really disgusting."

That really disgusting odor is caused by the offending bacteria, which expel two chemical compounds: putrescine and cadaverine.

"That's really what they're called?"

"Yup, it's kind of ridiculous," she said, "putrescine and cadaverine."

The scientists who named them could not have been less creative with their nomenclature if they had coined them grossicine and yuckicine. These two compounds are found not only in vaginas but also in spoiled fish and rotting bodies.

One study explained the scent as a "death-associated odor." A high-enough concentration of these compounds can be toxic. The vagina cannot produce enough to make anyone ill, though wouldn't that be a handy feature to have in our "Swiss Army snatch." Instead, the vagina can make just enough to forever link our private parts with fish tacos.

About twenty-one million women suffer from bacterial vaginosis, and the stink isn't the worst part of it; if untreated, it is also associated with pre-term birth, an increased risk of STD infection, and pelvic inflammatory disease. Some women seem prone to the condition, but the risk of getting it goes up for those who have multiple sex partners and/or use douche—two activities that tend to imbalance our bacteria. Douching is so awful, because douche amounts to staging a coup over your very close allies (the lactobacilli) and allowing those other stank-making evil bacterial soldiers to rule your southernmost cavity.

Because semen and blood decrease the acidity of the vagina, fishy smells can also occur temporarily after sex and during menstruation. Nothing to worry about there.

"You can't just stick yogurt up there to fix it, either," said Nelson,

A Time-out for Smegma Both men and women make smegma in their genitals. It's a mixture of dead skin cells and fatty oils. Smegma gets a bad rap for being disgusting, but the fresh stuff actually helps protect our goods and lubricate our parts during sex.

Smegma—a word that when said aloud can send the most composed among us into the fetal—greases the gears. For men, especially those who are uncircumcised, it enables the foreskin to slide away from the head of the penis without chafing. We women, on a smaller scale, enjoy the same benefit when our clits slip free of their hooding.

There is smegma backlash, because it's kind of like a dairy product—it has an expiration date. After a long day in pants or too long waiting in traffic, the smegma can build up and turn opaque. This white gunk stows itself away in our elegant and elaborately layered labial folds. At that point, it can develop a funk. To get rid of it, all it takes is a quick wash.

So smegma is many things—gross, helpful, terrifically malodorous, but most important, perfectly normal.

responding to a popular myth that the probiotic breakfast food that can make you (and Jamie Lee Curtis) "regular" can also help bacterially balance your lady parts. There are many different types of lactobacillus—the ones that live in your vagina are not of the same strain that lives in your top-shelf Greek yogurt. Nelson explained that putting yogurt up there would be like shoving Dobermans up your vagina when what you really need are pugs; the right species but the wrong breed.

"Studies have been done," she reported.

Sometimes BV will go away without any outside help, but it's best to visit the doctor for a proper diagnosis and a dose of antibiotics.

Before we got off the phone, I had one more question. "Can you change your smell by what you eat?"

I'd always been curious. I'd heard rumors for a number of years that there is a magical elevator in your stomach that brings pineapple and other sweet delicacies down to Vaginal Level for your lover to consume.

"So far, there is no evidence of that," she said, "but you can screw things up by smoking cigarettes."

She pretty much said that smoking cigarettes is like punching yourself in the vagina. It screws with your bacteria, making it more likely that you will stink down there—not like an ashtray, but like a salmon two days past its prime. (Yet another reason to switch to pot.)

The best thing to improve our pussy smell, counterintuitively, is to do absolutely nothing at all. Some water, sure, but you don't really need any soap. I'm positive most of us have heard this by now: "The vagina is its own ecosystem and the best thing to do is to leave it alone."

Personally, I prefer to think of my vagina as a self-sufficient adult who doesn't need my help—unlike me, she's got a job and knows how to clean up her own apartment.

When I think about this situation—that between my legs there lies a distinctive aroma—I'm reminded of an old philosophical question: If there's a

vagina in a forest and no one is around to sniff it, does it make a smell? That is another way to say, I didn't think twice about my vaginal scent until I found out it was something I was supposed to share with someone else.

In high school, I was blown away when I learned that other people, people I liked romantically, were supposed to enjoy diving nose-first into the depths of my loins. It's not that I thought mine smelled bad (I didn't even really take the time to investigate it all that thoroughly); it's more that I thought all crotches must smell bad. They are the Mariana Trench of the body, the deepest and most mysterious crevasse, after all.

From basic sex ed, I knew that oral sex was not going to make a baby. From basic evolutionary science, I knew that all animals' foremost goal was to procreate. Therefore, I developed my own theory about vaginal odors: Maybe it was actually the body's way of warning a dude that he was approaching the hole from the wrong direction.

I'm not sure where my insecurity came from, but having a boyfriend sniff my vulva made no more sense to me than telling him to make out with my shoes. I wasn't worried about him, but rather about what he'd think of me after being exposed to my most intimate and possibly (you never quite know what a prospective mate is going to be into) formidable fumes.

Because of this, for the first decade that I was sexually active, I did not experience one face in my crotch. I made sure of that. I used my thighs to clamp onto and halt any midsection that dared move southward while making out. I'd like to report that I was constantly turning eager men away, so much that I built up Arnold Schwarzenegger quads, but that wouldn't be true, so I can't. Of the men who tried, though, I appreciated their attempts to pleasure me even though it may have been confusing for them when I went full-on WWE.

During that time in my life, I often wondered why my clitoris couldn't be located more conveniently—why not, for example, near the very accessible, arid, and odorless elbow?

To get over my fear of having my vulva go face-to-face (so to speak) with a face, I went a bit extreme. (Though it made a lot of sense at the time.) I recruited a sex surrogate, a person who is trained to address sexual and intimacy issues via physical means, to sniff me.

For one, I figured that since he was a professional, out of integrity, he would tell me the truth about my odor. Also, I liked that no emotions were involved—it would be very liberating not to care about whether or not this guy, after experiencing my aroma, would call me the next day.

His name was Eric Amaranth and I went over to his home one winter evening. We talked for quite a while—he wanted to make sure I was comfortable—before I took off my pants. I lay down with my underwear still on and braced myself. He asked me if I was ready, and as soon as I gave him the go-ahead, he clinically stuck his large proboscis between my legs and took a couple of expert whiffs. When he reported back, he smiled and then likened my scent to a "twinkle." I felt like I'd just won on *The Price Is Right*, and if there had been a studio audience, I would have run up and down the aisles high-fiving everyone. I put my pants back on and left.

It was only later that I realized "twinkle" was not actually a scent. It meant about as much as saying my armpits smell like a "thump" or a "crackle." It didn't matter, though, because by allowing him to smell me and realizing that it did not cause the heavens to open up and rain frogs and leeches, I had already broken through the cunnilingus barrier.

I planned to spend the next five years making up for lost time by sitting on everyone's face, but only a few months after that great first huff, I met Dave, the man who would become my husband.

In the years since, I've become more comfortable. I've come around on my own scent—I play it cool when pockets of myself waft toward me in downward-facing dog—but I am still not completely liberated when it involves someone else. I happily allow visitations, but never without being a bit of a freak show.

When Dave and I are together, I make him reassure me that I smell like 100 percent fresh and pasteurized vagina. We'll be in the middle of something and then I'll pop my head up and say, "Everything okay? Are you okay?" He has to stop what he's doing and either say, "Yup," or give me a thumbs-up.

I've convinced myself that I don't like oral as much as finger stimulation, but it's entirely possible that's because I don't want to have to deal with feeling vulnerable. Then I feel shame about having shame. This is an era of female empowerment; it's so '80s to care about what your vagina smells like. I should be selling my dirty underwear for coin, dammit.

I'm not the only one with conflicted feelings. I was recently talking to my friend Jenny, who said that she was ashamed that she didn't like it when a guy went down on her and then tried to kiss her afterward. "Do you think that's symbolic of me not loving myself?" she asked.

I wondered if it meant that as well, but I felt like being supportive. "No," I said. "It means you have boundaries and that's totally cool."

The odd thing about all this is that I have no idea where this insecurity came from. I can't trace it back to anything specific—no one ever told me that I reeked, and my parents were very open and over-the-top sex positive. My mom was so down and open about the crotch that she would even put lube on the family's grocery list. I did have a male friend complain that he was with a woman who smelled like fetid two-week-old tilapia, which I'm sure added fuel to my fire, but I can't blame my insecurity on him because I was concerned long before I heard his story.

To get some context and hopefully come to understand how these conflicted feelings developed, I decided to look into the history of women's relationship to vaginal odor. I had a feeling that, as with most body shame, it stemmed from some bullshit buried in our cultural history.

Unfortunately, no museum exists with a wing dedicated to the Preservation of Pussy Odor Perspectives from Antiquity. For the next best thing, I called Karen Harris and Lori Caskey-Sigety, coauthors of the 2014 book *The Medieval Vagina: An Historical and Hysterical Look at All Things Vaginal During the Middle Ages*. My idea was to go back and try to pinpoint when women began to second-guess their own odor.

Quickly, I found that medieval times was not that time, rather it was more of a renaissance for vaginal funk. "After two years of research, we have come to the conclusion," said Harris, "that vaginal odor was not a priority for medieval women." According to the authors, women were cool with their own personal scent. They had bigger things to worry about, like the plague.

The authors suspect that in the Middle Ages a strong, gamy female scent was sexy to a man, a delicacy like oysters. "Or maybe they were just nose-blind," said Caskey-Sigety. "You have to understand, the ambient smells were rancid—spoiled food, raw sewage, trash, and a plethora of body odors."

"A vagina was probably a wonderful respite," said Harris.

I felt like the authors had just given me a pro sex tip: Next time I really want to impress my husband with my vaginal scent, I'm going to make him perform oral near a landfill.

"And they didn't get to bathe that often, right?" I asked.

"Yeah, there are some records of people only bathing once in their life."

The women during this era did douche, but the liquids were not used to "improve" scent. Instead, the mixtures were used to prevent pregnancies and disease. As Harris so cringingly explained, it was also used to "wash out the pus from their STD-ridden sexual partners."

Totally sexy, no?

One favorite douche of the era was made up of garlic and wine. "Wine-filled douchebags were probably the number-one go-to douchebag," said Caskey-Sigety. These ancient ladies were impressive: They were willing to share booze with a body part that couldn't even appreciate the flavor.

Though wine was often favored by women, gynecologists from that era—a.k.a. men who believed that the vagina was just a penis turned inside out—recommended that women stave off infections by squirting acacia, olive oil, pomegranate pulp, tobacco juice, honey, and ginger inside their genitals.

Coincidentally, that recipe doesn't sound too different from the fifteen-dollar "rejuvenation smoothie" from the juice place down the street.

In comparison with ancient Egyptian contraception, what the medieval ladies experienced was practically paradisiacal. Egyptians were instructed to stuff their canal with alligator dung. It is nice to think that there was once a whole nation of people who got excited when their lover brought home feces. Instead of slipping a condom out of his wallet, he would wink and then reveal a palm full of crap.

In any case, if you're consciously putting poop in your vagina, you are probably not overly concerned about how you're going to smell.

I was not able to find any record that throughout the next several centuries women felt self-conscious about their vaginal odors—no papyrus scrolls full of hieroglyphics lamenting the gases coming from underneath the female toga nor any ancient manuscripts directing women that "thy nether regions better smell like the flowery armpits of a wee newborn cupid." The Boston Tea Party happened, then the Battle of Bunker Hill, people traveled the Oregon Trail—throughout all that turmoil, nothing suggests that women thought twice about the scent that wafted from their Bermuda triangles. But also, be aware, we are sorely lacking in vaginal-odor historians.

The next big milestone for vag odor I came across didn't occur until 1832. An American physician named Charles Knowlton began promoting his own douching mixture as a form of birth control. The douche was a water-based solution that women spritzed up their canal after a hot night of loving; it included salt, vinegar, liquid chloride, zinc sulfite, and aluminum potassium sulfite. Clearly, Knowlton just dumped his leftover lab chemicals into a vagina.

Though not very effective, douching became popular for the same reason the Chicken McNugget, though not very delicious, became a favorite—it was cheap.

Now, as far as I can tell, this is where it starts to get bad for pussy odors. In 1873, Congress got uptight and passed the Comstock Law, making it illegal to use the U.S. mail to disseminate any information or paraphernalia regarding "erotica, contraceptives, abortifacients, or sex toys."

Douching companies, looking for a loophole in the new law, found a sneaky way around the censorship. They began to market birth control by rebranding it under a new label: "feminine hygiene."

In *Devices and Desires: A History of Contraceptives in America*, Andrea Tone explained that though douche executives could no longer claim that their products had contraceptive value, they implied it by using ambiguous language in their advertisements.

The ads freaked out young fertile ladies with headlines like "No Wonder Many Wives Fade Quickly with This Recurrent Fear," "Can a Married Woman Ever Feel Safe?" and "The Fear That 'Blights' Romance and Ages Women Prematurely." It was pre-internet clickbait. If I didn't know the context, I would think these were teasers for a horror film about uterus-eating goblins.

In the mid–twentieth century, Lysol became the leading pregnancy-preventative douche. When we use Lysol to scrub the bathroom floors today, we often wear gloves, but back in the Comstock days, women put the mixture straight up their vaginas. I had no idea that you could do that and not immediately die. Lysol ads claimed to be good on "sensitive female tissues," but they also urged people to use the same antiseptic liquid as a gargle, nasal spray, household cleaner, and dressing for burns.

"After disinfecting themselves, women could use Lysol to clean the garbage pail and toilet bowl," wrote Tone.

A competing company stated that its douche could be used both for "successful womanhood" and for athlete's foot. When things sound too good to be true—e.g., birth control that also cures an external fungal infection—they usually are. I'm not sure how effective the liquid was at cleaning the toilet seat, but as a method of birth control, these douches were worse than the pullout method.

In a 1933 study conducted at Newark's Maternal Health Center, it was found that 250 out of 507 women became pregnant despite using Lysol douche. When I learned that women still got pregnant while using that stuff, I had much more respect for sperm.

The douche, finally, was deemed a horrible and useless form of birth control, and yet it still didn't go away because douche, it turns out, is like the villain who doesn't die in a superhero movie. From what I could gather, it seems that when other, superior contraceptive methods such as the pill came onto the market, douche advertisements began to shift—touting it instead as a product with no other purpose than to make our crotches not actually smell like crotches anymore.

Douche, in other words, pivoted into a new market, focusing now on the noblest of misogynistic causes: to make women feel self-conscious and stinky. Here is a nice sampling of ads from the era: "Unfortunately, the trickiest deodorant problem a girl has *isn't* under her pretty little arms." "Why does she spend the evenings alone?" "The world's costliest perfumes are worthless—unless you're sure of your own natural fragrance." Then there were the notorious Massengill commercials that depicted mother-and-daughter walks on the beach—the duo look out at the briny ocean full of fish, while they claim to not feel quite right in the groin. The douching companies waged a crotchtacular smear campaign and it seems, to some extent, to have worked. In less than a century, vaginas went from fragrant to not the right kind of fragrant.

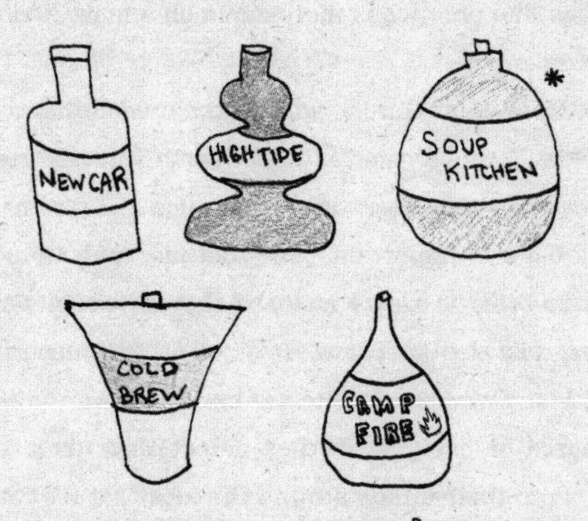

When I pass the feminine hygiene aisle at the drugstore, where boxes of douche are lined up on the bottom shelf just below the pregnancy tests and vaginal-itching creams, I often wonder why companies don't try to corner the male demographic with a parallel product for their balls. "Don't end the night by scaring her off with your scrotum." "Do you want her to give your balls bitch face?" "Size really doesn't matter if she can't breathe when you pull your zipper down."

Sheri Winston, the author of two books about women's sexuality, wasn't surprised by this disparity at all. She, in fact, believes that denigrating women's bodies is even more deeply rooted in our society. "Of course fear and loathing and negativity are attached to our earthy, musky, wet, you know, fertile fecund vaginas," she said. "In western culture, anything associated with women's sexuality has been vilified."

Douching fell out of favor when researchers discovered that the concoctions wreak havoc on our vaginal health, yet the Centers for Disease Control's National Survey of Family Growth found that among women aged fifteen to forty-four in the United States, about one in five still douches today. The practice is more common among African American and Latina women.

A nurse, Irene Raju-Garcia, who works predominantly with Latina women in New York City and is half Puerto Rican herself, explained that the women she encounters douche because it's ingrained in their culture. "Sex is big and important," she told me, "and women are raised to think that in order to keep a husband they have to not only look good physically, but they also can't have any smell other than perfume, flowers, or so forth." Many women come to see her when they have the slightest bit of discharge or even when they detect their own natural scent. "They can't wrap their minds around the idea that it's totally normal," she explained. Lemisol, she told me, is a favorite douche brand for the demographic.

I looked up the "feminine wash" online and found a bundle of rave reviews. One that stuck out was from Danielle M., who wrote that her "papi" turned her onto the product: "He said it would keep the taco smell out of my burrito and he was right. It brings all the boys to the yard and repels mosquitoes. Double win!"

Despite the few remaining douche loyalists, it is now much more common to come across a person who *is* a douchebag than the kind of douchebag that was made to hold vaginal douching liquids.

To find out how the douchebag made yet another transition—how it evolved into an insulting term for obnoxious assholes who manspread on the subway or rev their Harleys too loud—I contacted Taylor Jones, a linguist at the University of Pennsylvania.

"A lot of languages will use female genitalia to build insults," he said, "and the 'douchebag' has all the right components—the association with female genitals, bodily effluvia, and uncleanliness."

"So you're saying it's a shortcut for calling someone a dirty pussy?" I said.

"Yeah, something like that."

He explained that people have the same impulse in India. "A common insult there is calling someone 'a crispy snack fried in vulva sweat,'" he said.

I didn't argue, but that sounded like it could also be a term of endearment. "So does this happen because language is inherently misogynistic?" I asked.

"I don't know that I want to make too many statements about that," he said, "but historically in English and cross-linguistically, yeah, that's kind of a thing."

A profanity enthusiast, Reinhold Aman, over email, gave me pretty much the same breakdown. "Even if a douchebag is brand-new or clean, there's the underlying association with a dirty, smelly pussy," he wrote. "Thus 'douchebag' is used as a term of abuse." He was also sure to mention that *douchebag* is rather tame when compared with the worst malediction he'd ever heard, which he called "a shocking combination of blasphemy, scatology, and anal intercourse." The Hungarian lamentation he spoke of could be translated as follows: "Oh God, stop slapping me in the face with your cock all covered with shit from fucking Jesus." Reinhold was right; after I heard that, *douchebag* never sounded more like a fresh spring morning.

It was interesting, though—in a sense, each time we utter "d-bag," as fun as it is to say, we are, in an abstract way, proliferating the idea that pussies are dirty and gross.

After I'd looked at the history, it seemed like the fact that multimillion-dollar companies waged an all-out war against our vaginal odors was significant, but I wondered if that was enough to change one's perceptions of her own scent.

I knew I wasn't alone in the insecurity I'd experienced, after all. Many women have concerns that their crotches don't taste or smell like fresh-pressed unicorns. More than one gynecologist I spoke to revealed that she'd experienced a patient coming in after having poured Clorox bleach into her genitals.

In a *New York Times* article, the journalist Seth Stephens-Davidowitz crunched data culled from Google search engines. He found that while men were most concerned about their penis size (surprise!), women, when it comes to their vaginas, were most concerned about odor.

"Women are most frequently concerned that their vaginas smell like fish, followed by vinegar, onions, ammonia, garlic, cheese, body odor, urine, bread, bleach, feces, sweat, metal, feet, garbage and rotten meat," he wrote. In other words, many women believe that between their legs, they are baking a casserole from hell.

That's when I spoke to Rachel Herz, a scientist who has spent her career studying our olfactory sense. I probably wouldn't have believed what she said if she wasn't a professor at Brown University who had published more than forty articles in high-ranking science journals and written two books, including *The Scent of Desire,* about our sense of smell, but she explained that when we are born, our sense of smell is a blank slate, meaning we have no inherent likes or dislikes. She said that we learn the quality of scents both through personal experience and by what we are taught.

"Think about infants who have no problem playing with their poop," she said. "They have to learn it is not good, and once they learn that, the distinction becomes hardwired."

It was very hard for me to be convinced, but she said that something even as seemingly heinous as the smell of vomit could be appreciated for its complex sour notes simply if someone had taught me from infanthood that it was exquisite stuff.

Two parts of our brain, she said, are responsible for helping us acquire scent associations—the amygdala and the hippocampus. They are the first two relay stations to receive olfactory information, and those two areas of the brain process emotion and associative learning.

"As soon as we smell something," she said, "it becomes emotionally assigned to the experience we are having right at that moment."

When we smell the scent again, those associations are reignited. Our feelings about a scent will differ from the cultural norm in two cases: Either we have a personal experience prior to learning the cultural connotation or we have a personal experience that is so powerful it overrides the cultural connotation.

Herz told me the former happened to her in the case of a skunk. The first time she encountered the animal, her mother told her that it was a lovely aroma, and because of that, she attached a positive connotation to the scent. She didn't realize that was a culturally unfortunate thing to confess until several years later when she told her friends and was ridiculed. "If I didn't have that experience with my mother," she explained, "I would have acquired the dislike response."

Herz also gave the example of a woman who hated the smell of roses. Even though roses are culturally a very positive smell, this woman's first experience of the scent was at her mother's funeral. "That smell to her will always be very unpleasant and only associated with a sad, traumatic event," said Herz.

"So if we grow up hearing derogatory comments about vaginas or we see douche commercials that tell us we don't smell fresh," I asked, "is that enough to make us believe that the smell *is* bad?"

"The answer to that is absolutely yes," she said. She explained that being told that our vaginas smell like fish might even make us perceive the smell of fish when that scent is not actually present. "It's hard to overcome the connotation even if there is no actual connection to the slur," she said.

We also assign meanings to an odor, even before we've smelled the scent. If we hear vaginas smell bad, we are likely to perceive it as so once we put our noses down there.

Still, I was having a tough time getting this all through my head. "So you're saying that really, we have no inherent feelings about body odor?"

"There is historical evidence that deodorants of all kinds evolved because manufacturers wanted to sell people a product," she said. "It wasn't humans who had a predisposition that was negative toward their own odors."

Maybe that explained the disconnect; while many of us are concerned about our odor, there exists an actual market for women's dirty underwear. Vaginal scent is often vilified, but it is also put up on a pedestal. Companies tell us through subtle ads to get rid of our scent—"Be fresh"—while heteronormative masculine men are expected to profess a deep and abiding allegiance to ripe vaginas. It's all very confusing.

When I had spoken with Sheri Winston, the sexpert, she was absolutely adamant that our vulvas expel fumes that act as an aphrodisiac. She told me that our crotches smell for the same reasons a flower does—to lure our pollinators. "It's our own distinctive smell designed by Mother Nature to attract mates," she said.

Besides being a little skeptical of her metaphor—I've never seen a flock of horny people swarm around a woman with intense vaginal odor—I thought Winston brought up an interesting point. Is vaginal odor just an inescapable facet of being human, like tartar buildup on our teeth, or does it play a special role in getting laid?

I'd heard of sex pheromones and the like, but wasn't sure of what was true, so I went forth to see what science may have found on this front.

Very little, it turns out.

"There haven't been many studies about vaginal odor as an attractant," said Margo Adler, a researcher at the Evolution and Ecology Research Centre at the University of New South Wales. Apparently, it is difficult to find women who are willing to drop off their vaginal secretions every morning at a lab. It is also apparently taboo to design a study in which a group of naked women get sniffed one by one by blindfolded men as they rate their corresponding horniness on a scale of 1 to 10.

I spoke with George Preti again, too. He's the scientist from Monell Chemical Senses Center who'd tried to explain to me why Chucho, my dog, kept ending up with her snout in my groin. Early in his career, he also spent many years investigating female secretions. His ultimate goal at the time was to find pheromones. Pheromones are chemical compounds that some animals release—often in the form of oils or sweat—which other animals of their kind respond to physiologically. In essence, it's a type of chemical communication. Really cool sex pheromones have been found in other species like the male pig. The boar releases a substance called androstenone, which causes a sow in heat to arch her back and present her lady parts. I wish I emitted marastenone that would make Dave rub my back and do the dishes. This stuff works so well that it's been bottled and sold as Boarmate to aid pig farmers in artificial insemination.

So far, in humans, nothing so conspicuous has been found, but that is not to say that pheromones for sure don't exist in our genitalia. "We don't know yet," said Preti, "a lot more work needs to be done in this field."

Then Preti had a small but I'm sure totally valid rant about the difficulty of getting funding in science these days. The funding is just not there for vaginas.

Despite the sparse research, there were some academic researchers who were willing to entertain my questions and take an educated guess. One was Gordon Gallup, an evolutionary psychologist who hails from the University of Albany. He's known for his out-of-the-box theories about body parts. He developed the "semen displacement theory," which explains why the human penis looks like it's wearing a Darth Vader helmet. He believes this distinctive ridge is an inverted shovel, made to scoop the sperm from sexual rivals out of the vaginal canal before a man shoots his own load inside. I'll never look at Darth the same.

"Do you think vag odor can help attract a mate?" I asked him.

Gallup does not look at evolution as survival of the fittest: "It doesn't matter whether you live or die"—it's whether or not you pass on your genes before you drop dead. He therefore looks at vaginal odor from the perspective of how it might help us successfully reproduce.

He referred to one of the two studies conducted about vaginal odor as an attractant. Both were conducted way back in the '70s, and because the sample size was so small—only four women's vaginas—some scientists, like Adler, dismiss these studies altogether. Nonetheless, the researchers found that vaginal odors, in the parlance of the study, were "less unpleasant" during ovulation.

"Males tend to rate vaginal odor samples that are taken during the fertile phase of the menstrual cycle as being more attractive," said Gallup. "Isn't that interesting?"

Some animals, like kangaroos, on the other hand, find out when a potential mate is ovulating by taking intermittent sips of her urine. If you ever thought it would be fun to be a male kangaroo, now might be the moment to adjust that opinion.

"So basically, when you boil it all down," I said, "vaginal odor is more pleasant during the ovulation stage, which might make the man and woman more likely to mate during that period and therefore be more likely to conceive a baby."

"That's exactly right," he said, "but I think body-odor cues had much more salient properties and much more profound effects during human evolutionary history than they do now."

Now we deodorize and lock up all that pubic potpourri inside our pants. A flower, in other words, though fragrant and beautiful, can't attract a bee if it's all boxed up.

Saying that vaginas smelled better during ovulation was not the same thing as saying that men were actually attracted to the scent. Because of the lack of research, I decided to conduct my own small study. I wrote a post on both the Los Angeles and New York City Craigslists asking men if they would be open to being interviewed about women's vaginal odors. I kept refreshing my in-box, but did not get one response. The next day, I tried arguably the most open-minded demographic in the United States when I posted the same question to the San Francisco Craiglist's "casual encounters" page. I knew it was biased from the get-go, like doing a survey of fashion and using respondents only from Milan or one about the benefits of juggling and querying only professional clowns.

I could have just asked my husband or culled evidence from past relationships, but clearly, I wanted fresh, unbiased discourse with men who were stalking casual sex on the internet. Before my post was flagged and removed two hours later, I got fifty-three responses. I spoke to six of the men.

I was expecting to find 100 percent admiration of vaginal scents—dudes who were connoisseurs and could admire even the deep funk of a crotch that has been stuck in jeans during the entirety of a cross-Atlantic flight—but in actuality, I found a range of sentiments. Most of the men were deeply enamored by the smell, but only if it was from a "clean vagina."

For example, I spoke with Julius, a sixty-year-old with a thick European accent. I was trying to keep him on track, but it was hard. "So you're saying you don't like it when a woman has little toilet paper balls stuck in her vagina?" I said.

Julius was obsessed with hygiene. We were already twenty minutes in and still talking about the toilet.

"Yes, I don't like that," he said.

"But toilet paper balls are a whole different story from the vaginal scent," I explained for the fourth time.

Julius said that he wished America would adopt use of the bidet, because he found the true essence of the vagina quite pleasing, but it was hard to tap into the real thing without a lot of prewashing. He spent the next two days sending me links to his favorite bidet models.

There was also Adib, a twenty-something who enjoyed picking women up at clubs. "If you had a stinky pussy," he told me, "I wouldn't fuck you."

Unfortunately, I lost my journalistic objectivity for a moment and took what he said personally. "Well, I wouldn't fuck you at all. Ever!"

"Lady, chill," he said. "I wasn't talking about you specifically."

His perspective just seemed so superficial, and I felt a pang of distress for fragrant females everywhere, but I tried to reel it in and regain my cool enough to continue interrogating him. "Right, sorry," I said, trying to get it together. "So what does 'stinky' mean to you, exactly?"

"That it's stinky," he so graciously described.

On the other side of the spectrum was Michael, whose favorite thing in life is to wear vaginal secretions all over his face. He won't wash up after going down on a woman because, he swears, the fragrant liquid gives him a superpower. "I get treated better when it's all over my face," Michael explained.

"Really?" I asked.

"Oh yeah," he said, "you should see the level of service I get at restaurants."

Then there was Adam, who was a little too woo-woo for me. "When I smell and taste a woman," he said, "I know exactly how her day was and how she's feeling." Don't get me wrong—I liked the idea of vagina as crystal ball, I just didn't buy it.

Then there was Jaime, who kept changing the topic—odor—so that he could tell me over and over again how he makes girls squirt. "She was embarrassed about it," he said, "but I kept telling her that her sacred waters were beautiful."

There were such wide-ranging opinions on vaginal aromas that I couldn't draw a significant conclusion. In essence, my experiment kind of backfired: Instead of learning about vag odor, mostly I had nightmarish memories from my dating days.

Sometimes you can't truly understand what you have until you don't have it anymore. In that spirit, I decided to do one more experiment. I know douching is horrible for you: You are committing genocide against your unassuming bacterial allies—WARNING: DON'T DO THIS AT HOME—but this seemed like the only way to understand the role vaginal odors play in modern daily life. Would my husband and dog change their behaviors if I didn't smell like vagina anymore? Would my dog think that I was a stranger? My husband get hornier if I smelled like air freshener? Would he even notice?

I told Dr. Jenny Hackforth-Jones, the ob-gyn I had spoken to earlier, that I was going to douche. I wanted to get her read on how dangerous it was. “Nooooo, don’t do it!” she said. “You can buy it, take the stuff out of the box, and look at it, but don’t do it.”

“I’m just going to do it once,” I said.

“You’ll get horrific vaginosis!” she warned.

I told her that I had to take the risk. This was what I could do for humankind. Some people save the whales. I douche in order to contribute to the comprehension of vag smells.

I went to a Walgreens where I don’t usually shop. My customary cashier at the closer Walgreens already knows that I need witch hazel pads for my hemorrhoids and a cheese grater–like contraption for my toe calluses. I believe in spreading out my issues among various drugstores. I don’t want any one clerk getting too much of a sense of me and my dysfunctions as a whole.

I walked through the fluorescent-lit store until I reached the feminine-hygiene aisle. I didn’t want anyone to see me—not because I was afraid they’d think I had a smelly vagina, but because I didn’t want them to think I wasn’t liberated enough to handle it. I crouched down and saw that there were a surprising number of douches to choose from. Ultimately, I went with a classic, Sweet Romance from Summer’s Eve. The pink polka dots on the packaging spoke to me. I turned the box around and read the warning, “An association has been reported between douching and pelvic inflammatory disease (PID), ectopic pregnancy, and infertility.”

Perfect.

When I got home, I brought the douche kit into the bathroom. Dave was due home from a short business trip later that evening. My dog, meanwhile, sat on the furry white bath mat. As I took the squeeze bottle out of its packaging, I projected concern onto her little face—wide eyes with her brows bunched together. Presumably, I was about to wipe out one of her favorite pastimes. In a few minutes—if all went as expected—my private parts were going to smell like a Strawberry Shortcake scratch-and-sniff storybook.

Douching was way easier than I thought it would be. It was like squirting mayonnaise onto a sandwich; only the sandwich is actually your vagina. Douche companies make it simple by attaching a nozzle to a disposable transparent bottle preloaded with the liquid. The directions said that the liquid should "flow freely out of vagina." I'm glad they said that because, as a douche novice, I would have thought you'd leave it in there for a bit to slosh around. I wanted to hate every part of this backward practice, but the sensation—drinking an ice-cold beverage from the wrong end—wasn't entirely loathable.

I toweled off, stepped out of the tub, and immediately tried to sniff myself, which is always complicated because, well, ribs. I was concerned about how the perfume would mix with my natural odors. Think about it: No one is fooled when a match is deployed to overcome a particularly fierce bathroom episode. Rather, they combine to make their own unique miasma.

After a substantial time spent contorting and getting my paws down there, I still didn't catch a whiff. After a few hours, I resolved that my smell—rather than transforming—had actually been canceled out. Oddly, I smelled like nothing at all.

Despite the fact that I'd just willingly done something that might cause a "death-associated odor" to develop in my groin, I felt pretty average. I slipped on some pajama bottoms and lounged around the house, waiting to see if my dog would notice anything different. Not much later, she trotted by and sniffed me. I can almost swear she did it with less finesse and joy than was typical, but my vaginal myopia might have been biasing me.

When Dave arrived home several hours later, he did not stop everything at the threshold and say, "Something is amiss—I can't smell my wife's genitals!"

The next day, he didn't notice anything, either, but he probably shouldn't have, given that I had pants on most of the day. The day after that, we were together and while we were still in bed, I said, "Notice anything different?"

He got that oh-shit-I-was-supposed-to-notice-that-something-was-different look on his face.

"I douched!" I said.

"I guess there was maybe something a little different," he reluctantly agreed.

"Did you like it better?" I asked. "Did it turn you on?"

This time he played the diplomat. "I like you both ways," he said.

I got a little defensive. I wanted him to say, "Babe, it's always better when you're funky." So I said, "Don't you like it better when I'm funky?"

He just sort of shrugged. He was not at all invested in this experiment. It was as if vaginal odor did not even take up the tiniest slice of his pie chart.

Over the next few days, I found that vaginal scent was important not only because it serves as the best indicator of whether or not that random pair of underwear hanging around is clean or dirty (it was a confusing week for laundry), but also because of something deeper. Back when I talked to Dr. Jenny, she told me that when women reach menopause, their vaginal smell changes because of hormonal shifts. "They are always surprised," she said, "that their scent is something they miss."

At the time, that reaction seemed incomprehensible, but after a few days, I understood. With my eyes, I can see my reflection in the mirror, but my scent is another kind of self-portrait. When reflected by a nose, it's just as vivid and familiar. Without my normal scent—a thing I didn't even realize I registered on a daily basis—a part of me felt foreign.

I thought my odor would have the most profound impact on my husband and dog, but it was on me—I actually missed myself.

A Field Guide to Vaginal Discharge

Discharge cleans the vagina in a similar way to how saliva cleans the mouth—the fluid flushes out dead cells and bacteria. In a healthy vagina, the discharge is white or clear and varies in quantity and texture throughout the cycle. There is often a spike in production just prior to ovulation. During that fertile time of the month, the discharge becomes clear and viscous like raw egg white. If the discharge is elastic—it can stretch without breaking, like hot pizza cheese—it is said to have "good *Spinnbarkeit*." What a wonderful world—there exists a word with the sole function to describe the stretchy properties of cervical mucus.

The amount of discharge can also vary from woman to woman. One woman told me that her underwear, at the end of each day, is akin to a thrice-used Kleenex. I did not resonate with that metaphor. When I heard that, I was concerned that I didn't discharge enough. Am I a good discharger? Do I discharge right? We all, it turns out, discharge different.

Discharge isn't there only to clean, lubricate, and protect; it is also the vagina's way of sending out an SOS signal. If something is amiss, the discharge may change color, consistency, and/or odor. Discharge, then, can also be the vagina's way of etching H-E-L-P into its desert-island underwear.

This field guide to vaginal discharge is inspired by the long-running practice of comparing women's body parts (and the size of their fetuses) to edibles—melons, buns, tacos, clams, etc.

COLOR: Thick and white
FOOD DOPPELGÄNGER: Cottage cheese
ODOR: Low
CAUSE: Yeast infection
PAIRS WITH: Cotton underwear

COLOR: Brown
FOOD DOPPELGÄNGER: Bone broth in its chilled gelatinous form
ODOR: Mild
CAUSE: Though there are other reasons, it is most common at the end of menstruation.
PAIRS WITH: Pizza and Netflix

continues →

COLOR: Green and frothy (but sometimes yellow)
FOOD DOPPELGÄNGER: Matcha latte foam
ODOR: High
CAUSE: Trichomoniasis, an STD that causes itching and burning
PAIRS WITH: An awkward conversation

COLOR: Grayish white
FOOD DOPPELGÄNGER: Mushroom soup
ODOR: High
CAUSE: Bacterial vaginosis*
PAIRS WITH: A round of antibiotics

**Bacterial vaginosis is diagnosed with a method called the Whiff Test, which is no more sophisticated than it sounds. The doctor takes a sample of your mushroom soup, mixes it with a few drops of potassium hydroxide, and then takes a big, brave whiff. If she smells fish, then you have BV.*

COLOR: Cloudy yellow
FOOD DOPPELGÄNGER: Lemon butter sauce
ODOR: Medium to high
CAUSE: Gonorrhea (or if you're feeling fun, the clap)
PAIRS WITH: Not pairing

The Butt Paradox

Butts are a paradox. They are where poop comes out, which is gross, yet they are also one of the most sexualized parts of the human body. When I see a snug pair of jeans on a nice round rear and say, "Look at that ass," it's really not very different from fawning over a garbage truck.

We enjoy the ass—put photos of them on the front of magazines, show them off in tiny bikini bottoms, and work hard at the gym to maintain them—despite the fact that those two cheeks seem to be little more than an overglorified welcoming committee, designated to stand still and steadfast while giving yesterday's dinner its last respects.

Don't get me wrong, I'm perfectly happy that people don't look at my behind and immediately think *sewage*, but how is it that a society so easily turned off by bodily functions is able to overlook the less savory bits of the rear?

To investigate, I first went to Los Angeles to speak to Cindy Thorin, a fifty-nine-year-old aesthetician with the warmth of a grandma mixed with the rawness of a drunk cast member on *The Real Housewives of New Jersey*. I thought she'd have some insight into the paradox because she is famous in anal-bleaching circles for her anal-bleaching creams. By erasing the pigmentation around the anus, she attempts to do for the butthole what an interior decorator does by hanging a fancy curtain around a water heater—hide the inner workings. Her salon, Pink Cheeks, was decorated like a cozy living room, replete with fireplace, and there were small signs for sale that said things like "Sweat is fat crying" and "When life gives you lemons, a simple operation can give you melons."

"Hello," Thorin said, bringing me into one of her treatment rooms, empty save for a waxing table covered in tissue paper. Even though she has spent her career trying to create the illusion that butts don't poop, it turned out after a brief conversation that she herself was mystified by why people go gaga over the body part. "I've worked with butts for thirty years and I don't get it," she said. "They don't do anything for me." She told me that she thinks many people actually bleach because they are misinformed about the pigmentation. "They think they didn't wipe well and got a stain." As she explained this fact, she shrugged. "I tell them it's just genetic. I say, 'Your grandpa or grandma probably had a dark one.'"

After talking to her, I didn't feel any more enlightened, but since I was already there, I had her assess my butthole. (When else would such an opportunity arise?) I took off my pants and then did as she instructed: got up on all fours on her waxing table. While she took a gander inside, I looked at a poster on the wall of a callipygian pin-up model.

"You'd be a good candidate," Thorin said. "You're pretty ashy."

I don't know why, but I'd had the feeling I came from dark-anus lineage.

When I left Thorin, I knew I had to look at this issue from a different angle. I soon got a tip from an evolutionary theorist; he told me to call up Melanie Shoup-Knox, a behavioral neuroscientist at James Madison University. When I posed the question to her, she gave me something I could work with; she told me that there's a lot more to butts than meets the consciousness.

Researchers have found that men are more attracted to a butt if it and the hips it hangs on are larger than the waist. Estrogen, which begins circulating during puberty, causes fat to collect on our backsides. "It is a cue to how fertile the female is," Shoup-Knox said.

Fertility isn't the only factor for being drawn to an ass; the fat that is stored on our hips and butt is also specialized. This fat—long-chain polyunsaturated fatty acid—is stored preferentially on our lady hindquarters. These fats, Shoup-Knox said, are reserved and recruited and burned only during the third trimester of a pregnancy and during postpartum breastfeeding.

"There has been a reevaluation of why men prefer big butts, and we believe that's because those are females who can produce the most intelligent offspring," she said. "Those special fats are baby brain food."

That made me the tiniest bit self-conscious. "So, like, how big does a butt have to be to make smart babies?" I asked. I have some flesh back there, but not nearly enough for Sir Mix-a-Lot to worry if I started doing side bends or sit-ups.

"Well, it's not so much about the size of the butt, but more about the ratio between the hips and the waist."

"So a small tush can still make a smart baby?"

"Exactly," she said, "as long as the waist is smaller than that butt."

She went on to tell me that extra-large asses, like Kim Kardashian's, won't up a kid's IQ any more than a typical butt. "There's a point when you stop getting returns," she explained.

According to this research, males who prefer that feature—a big, perky bottom (or to be more specific, the waist-to-hip ratio of 0.72)—will be more likely to choose a fertile mate and have offspring that are more intelligent. If his babies are more intelligent, they are more likely to survive, and therefore the dude who likes big butts gets his genes spread onward and outward into the horizon of human existence.

"It's not just the butt that's being selected for," Shoup-Knox said, "but the preference for the butt is also being continuously selected for."

In other words, as much as a big butt is passed down from mother to daughter, Shoup-Knox believes the attraction to big butts is also passed down through the generations.

"So this is enough to make people overlook the less pleasant aspects of the butt?" I asked.

"Reproduction is so essential to our survival that we can overlook anything if the reproductive benefit of that feature is *that* great," she said, "and in this case, it is."

We are all probably destined to look back on this big-butt science skeptically—isn't it a bit too convenient that everything always comes back to procreating, and how do we explain women's enjoyment of the male derriere, which will never yield "brain food" for our young?—but for now it's what we've got and Shoup-Knox was convincing. Even so, I wasn't sure she had the whole story.

Before concluding this investigation, I went directly to the source—the people who have butt affinities. First, I asked my friend Josh about his butt adoration. He spoke very broadly. "I like something to hold on to," he said. With a generic line like that, I decided he was just in poop denial and wanted smart babies.

But next, I visited my friend Mark, a worshipper of the female behind, and he shed some new light into a dark crevasse.

I gave it to him straight. "How do you maintain your love of the butt when we excrete from the middle?"

"I love a hole," Mark said. "People are obsessed with holes. I think somehow we are hardwired to like holes."

"I thought people liked the butt in spite of the hole," I said. "They overlook the hole."

Mark was adamant that the hole was actually the draw, and the rest—the cheeks—was just decoration like a frame is to a masterpiece. "I think the hole is attractive," he said. "And what makes it attractive? It's a hole!"

He went on about holes for so long my recorder almost ran out of space. "If you give a boy a little round doughnut-shaped toy," he continued, "he's going to stick his finger in the hole. We like holes. We like sticking things in holes!" Mark was getting too excited.

"I get it," I said. "You like holes."

Mark's poetics about the hole brought me to my final observation: pleasure. The butt is not only a two-cheeked vessel signaling fertility or a cushion on which to rest one's weary head; it also houses a very real erogenous zone. Many people can overlook a lot of heinous stuff—messy rooms, bad breath, the exit shoot of the intestines—in order to have orgasms.

Ultimately, there are so many positives—both conscious and subconscious—about the tush that these factors successfully undermine the reality: Our butts, even the firmest and perkiest ones, are full of crap. In other words, "I'd love to get my hands on your septic tank" is not a pickup line that will hit the streets anytime soon.

PILEup on thE "InneR" State

I'll always have one big regret in my life. There was a question I yearned to ask my grandfather before he passed. My mom had told me that he possessed great knowledge and experience on this particular issue, but each time I visited his home, I didn't have the gall to bring it up. When he got sick, I knew time was running out, yet I still couldn't muster the courage. Every time I attempted the conversation, my heart would beat hard and my face would grow hot and flushed.

I never managed to broach the subject. In August 2011, Irvine Harold Altman, my beloved grandfather, took eighty-nine years of wisdom on the matter to his grave. No reflections on the subject were entrusted in his will. I experienced deep remorse and helplessness: I felt alone in the world, left to find the secret of how to deal with a hemorrhoid flare-up on my own.

When I turned twenty-six, my butthole staged a mutiny. It felt like a gang of misfits lit a bonfire in my anus. Whenever I sat, I'd swear I plopped down right on top of a demon's trident. Wiping, that unexceptional and often forgettable activity, became a form of torture. Starving myself sounded like a great option; if I didn't eat, I wouldn't have to go to the bathroom. I might die, but that was okay!

Before this happened, I had been obsessed with all things fecal. I was that friend, the one you would text when you needed to debrief about that satisfying yet unsettlingly large dump you just took. This turn of events felt ironic, like a composer going deaf.

I didn't know what was wrong and yet I was too terrified to find out by looking in the mirror. What if my anus was the size of a basketball? All I knew was that the pain—deep and searing—felt metastatic and quite possibly terminal.

After weeks of shuffling around like a ninety-year-old man wearing an ill-fitting diaper, I went to the gyno under the guise of needing a Pap smear; I was too embarrassed to make an appointment solely for my butthole.

There was something odd about the ass—especially maladies of the ass—that rendered it unspeakable. In a booklet about constipation called *Let's Get Things Moving*, the authors note that when we are young, we are enthusiastically applauded for pooping, but simultaneously taught that it's dirty, smelly, and disgusting. "A very confusing state of affairs," they rightfully concluded.

After the swabbing, I reluctantly asked my gyno if she would look at what kind of nightmare was brewing a bit south.

It took her only a second to diagnose me. "Oh, you have a little hemorrhoid," she said. Her tone made it sound like she'd just discovered a cute kitten hiding in a box.

I must not have heard her right.

"A what?" I said.

"A hemorrhoid," she said, taking off her gloves and making a few notes on a clipboard. She told me it was a tiny lump, not much bigger than a mosquito bite, and explained it as nothing more than a swelled vein in my butt.

To her, this was no big deal, but I'd always thought that hemorrhoids happened only to obese truck drivers with ketchup-stained T-shirts and a lifetime subscription to *RoadKing*.

How could I have a hemorrhoid? I had flower-patterned underwear, for God's sakes. Wearing those dainty things with hemorrhoids would totally clash.

"Just wait until you're pregnant," my gyno said, with a sinister wink.

Hemorrhoids are extremely common, but I didn't know that for years. So for the time being, I kept the 'rhoid to myself. While most of my friends were reading love stories, I mined literature for hemorrhoid references.

I found solace in fictional characters who also suffered from the affliction, like Marx Marvelous from *Another Roadside Attraction*. "You women think you have it bad, having babies," he said in the Tom Robbins novel. "Well, a woman can only give birth once every nine months, but a hemorrhoid sufferer goes through labor every time he goes to the crapper."

Yes, I thought, Marx *gets* me.

I took baths to soothe the pain, bought supplies with "anti-hemorrhoidal" on the label, and complained to my mom. That's when she tried to get me to do an AMA (Ask Me Anything) with my grandpa about his own ailing rear. Obviously, that didn't go well.

My hemorrhoids, since then, have come and gone like migrating birds. One month they are here, the next they take off. Even though I've studied up on the 'rhoids, I've never fully understood them. The whole thing—how a body part could suddenly malfunction—has continually pissed me off. Plus, with an anus this sensitive, I'm not sure how I'll ever get to try butt sex.

I could draw only one conclusion: Our anuses weren't constructed correctly. The little muscles in our behinds have been finicky since we stepped out of sludge eons ago. Since ancient Egypt, scholars have been lodging formal complaints about anuses—a remedy for hemorrhoids was found written on papyrus from 1700 BC. Clearly, the anus should have been made out of a sturdier material—titanium perhaps, maybe concrete, even some nice hardy plastic would be cool.

HEMORRHOIDS ANONYMOUS

I contacted Peter Lunniss, a colorectal surgeon, to present him with my findings. I discovered him because of his copious writings—more than ninety publications—about defecation.

"I feel like our assholes weren't made right," I explained to him. "There are so many issues."

"No, it's just that we abuse them," he said.

"It's not them, it's us?"

"Yes, it's us," he said, "definitely us." He reiterated that our problems most certainly stem from user error and then went on to praise the anus for being one of the most complex organs in the body—surgeons can transplant hearts, livers, kidneys, and even faces, but they have yet to figure out how to reuse someone's old anus. "People just take their anus for granted until it goes wrong," he said. "You've got to look after them."

"So then what about hemorrhoids?" I said, expecting that would be enough to sway him to my side.

From what I'd read, 40 percent of the population has had or will have hemorrhoids at some point in their lives. Most studies even suspect the number is higher given the amount of people who are too embarrassed to tell anyone in the medical field about their wonky buttholes.

Lunniss told me that I'd been thinking about hemorrhoids all wrong. We are all actually born with them—everyone from a day-old baby to Beyoncé. They begin as anal cushions, three of them, that help our anus remain airtight. "You'd be leaking gas without them," he said.

He explained that they give our anus an extra seal, "like the rubber ring at the top of a jam jar."

Doctors seem to have no qualms about ruining the connotation of perfectly good kitchen items. I once had a gynecologist describe a speculum as being like tongs.

Lunniss said that for a variety of reasons, like pregnancy, being overweight, and constipation—anything that adds more pressure to our bottoms—the cushions may get pushed down lower and become congested with blood, which causes them to itch, hurt, and sometimes bleed. When the cushions get displaced in this way, they become their evil alter ego: hemorrhoids.

Hemorrhoids, then, are not some weird growths or, as I once suspected, a sign that everything you love and cherish in the world spites you. They are actually a part of our body that isn't working 100 percent at the moment.

You can get hemorrhoids lopped off with a scalpel, blasted with a laser, or stapled with, well, a stapler, but Lunniss said that it's best to use surgical options only as a last resort. "If it's not done well," he explained of the surgery, "you can lose that seal—know what I mean?"

"I think so." I must not have been convincing, because he continued and became even more explicit.

"You may no longer be able to hold wind," he told me. The best thing to do, he said, was to learn how to shit properly.

"I know how to shit properly," I said.

He wasn't so sure. "The civilized world is obsessed—they will force themselves to poo without having an urge to evacuate their bowels," he said. "If you're straining all the time, you're eventually going to end up with prolapsing hemorrhoids." He got really ominous; he said many of the young are careless with their butts and will see the consequences down the road.

For the next hour, Lunniss lifted the veil, enlightening me about the magnificence and complexity of the anus (and how we screw it up). If we step forward and then take a gander inside, the anus is much more complicated and sophisticated than we give it credit for.

I was astonished to learn that we have not one but two sphincters. The one many of us are more familiar with is the one that we can clench on command. This is the external sphincter, and we control it voluntarily.

The internal sphincter is the behind-the-scenes sphincter—you can't see it and can't control it, but without this sphincter, the show wouldn't go on. It is positioned inside the anal canal and is the size of the thickest string on a guitar. It is contracted at all times except when we have to go to the bathroom.

"So, what if we didn't have that internal sphincter?" I asked.

"Then we would have to rely on our external sphincter to keep the anus closed and it would get tired after about twenty-five seconds and then you would poo your pants," said Lunniss.

For two weeks, I told everyone I met, "Did you know that we have two anal sphincters?"

(The internal sphincter needs much better PR.)

Each sphincter answers to a different boss. The external sphincter pledges allegiance to the conscious bits of our brain. In medical circles, the urge to go to the bathroom is formally referred to as "the call to stool." Why they didn't name it "the call of dooty" we'll never know. When you feel "the call to stool" while you're on a ten-hour bus ride with absolutely no toilet in sight, it is the external sphincter that you will clench and pray to. It is the more civilized of the sphincters. It cares what people think of you.

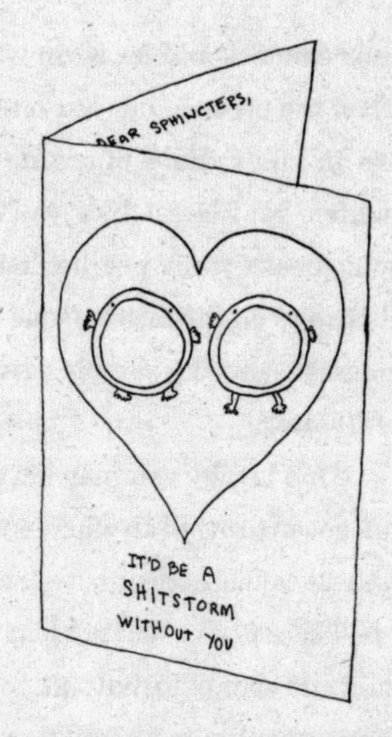

The internal sphincter is the much more practical one. We have as little say in what it does as we do in the beating of our own hearts. It does not care if you are at your new lover's apartment; it would happily have you wreck the toilet. If the sphincter had its way, a quiet library would have no power against the onslaught of your farts. This sphincter has absolutely no shame. Its main concern is to keep the rectum unpacked and to prevent backup.

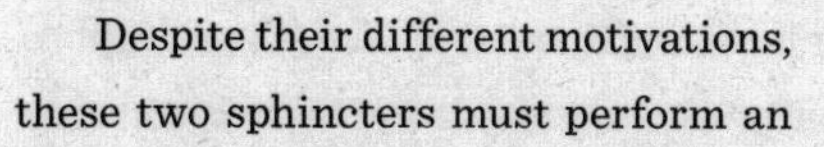

Despite their different motivations, these two sphincters must perform an intricate tango to remove waste from your body (and keep you looking cool while doing it). When digested food comes down the pipeline and hits the internal sphincter, it opens up automatically. The contents of your bowel don't just barrel through and crash against the external sphincter, hoping it will open. The internal sphincter is polite—at first—and will let only a tiny "sample" pass. This sample goes into the intersphincteric zone between the two sphincters—poop purgatory, if you will—and is analyzed by specialized sensor cells. These cells exist nowhere else in the body. Within an instant, these cells tell us exactly what's knocking at the gates—something liquid, gaseous, or solid.

Hearing about this might have been the closest to a come-to-Jesus moment I've ever had. "You're right," I said, "we almost always know if it's a poop or a fart!"

"The anus is so clever," said Lunniss.

Once the materials are analyzed, the external sphincter waits for us to survey our environment. If you sense something gaseous and you're at a loud and bustling farmers' market, you might choose to take your chances and let it rip while you quickly run off to the next booth. If, on the other hand, you realize it is something more than gas and you are in the middle of giving a talk to five hundred people, you will tighten your anus like you've got a winning lottery ticket between your cheeks while perched on the deck of a yacht going 70 knots. If you choose to do the latter—squeeze tighter—the internal sphincter gets the message, and the sample gets sucked back up, where it will be deferred until later.

This is why you may have experienced running toward the toilet at full speed to avoid an emergency only to find that once you've finally made it there, you no longer have the urge.

The reprieve will not last for long, though. There is a plane waiting to land and your external sphincter is air traffic control. The plane might go for another lap, but it will eventually need to come down. It's only a matter of time before the internal sphincter will give it another shot, sending another sample through. When it does, and if this time you're prepared—near a toilet, magazine in hand—you're golden. Both sphincters will relax. Bombs away.

Every time we go to the bathroom, I observed, it is a compromise between our animal and civilized sides. The internal sphincter is our internal pigeon: It thinks the whole world is its toilet. The external sphincter, meanwhile, is the city mayor who doesn't want pigeon scat all over every statue.

Lunniss said that one reason we get hemorrhoids is that we give our mayors too much power. In that way, anuses are like most governments. Give any elected official too much authority and something bad will happen. "A natural poo is when you listen to your body," he said. "That's what I'd call a good poo." He explained that far too many people strain these days—they want control of when and where they go—and it's breaking anuses everywhere.

"I've never had the patience for intestines," I admitted.

"Well, there you go," he said.

Many people, he continued, also refuse public restrooms in favor of the familiarity of their own four walls. In other words, they ignore their urge until *they* deem it is the "right time." I luckily do not have that problem—I have no special allegiance as long as there is toilet paper—but I have many friends who would prefer to be filmed in HD streaking across a packed football stadium than to drop their pants in a Target stall. While this might be good for the soul and the psyche, it is not good for the sphincters.

We can overpower the urge only so many times before we get into trouble. The internal sphincter will give up on us. I spoke with Theresa Porrett, a coloproctological nurse and consultant at London's Homerton University Hospital, and she gave me her best rendition of a dejected internal sphincter's thoughts: *"You're not going to listen to me anyway, so what's the point?"* In that way, the two sphincters have a relationship like many of us do with our loved ones. If we feel taken for granted or unappreciated, we may begin to stray.

A straying internal sphincter is bad. You really don't want your internal sphincter muscle to break up with you.

"Constipation is often the result," said Porrett.

"Why?" I asked. "How does that work?"

She said that if you defer too often, you will stop getting the urge, which means that your feces will sit in the rectum for longer. While it sits there, your body begins to reabsorb all its water. "It gets smaller and smaller and harder and harder," said Porrett. "You'll be stuck straining to push out tiny little rabbit pellets instead of a nice solid, soft sausage."

Porrett was another medical professional who seemed to have no qualms about ruining things: this time the connotation of perfectly good deli meat.

After a few moments, she said, "A soft sausage really is the best way to describe it."

Later I would speak to another nurse who described the resulting constipation a little differently: "You will have a brick in your bottom and good luck getting that out."

When people are constipated, they tend to strain, which, again, can piss off the old anal pads.

Porrett wrote a paper on the topic, titled "Take Care of Your Bowels and Your Piles Will Look After Themselves." (*Piles* being another word for hemorrhoids. It comes from the Latin *pila*, meaning "ball." I didn't think this topic could keep getting sexier, but it just did.)

Besides drinking lots of fluids to stave off constipation and eating breakfast in the morning to stimulate our intestines, she said another important factor in anus health is our position on the toilet. "It's important. Really important!"

She has it out for the Western toilet. "We design toilets to match the sink and look great with the shower," she said, "but we don't design them to actually help us poo better."

Maybe Porrett was onto something. My parents' peach toilet matches their peach sink, their peach tub, and even the flecks of peach in their tile.

There are two basic schools of thought: squatting and sitting. Squatting was the go-to since just about forever, while sitting became popular only with the advent of indoor plumbing in the 1800s.

Sitting, while popular, wreaks havoc on our buttholes. Porrett explained that because we stand on two legs, there is a lot of pressure on our lower region, so several fail-safes have been put into place to ensure all our goods don't fall out when we stand up or sit down. Our sphincters are one fail-safe, but that's not enough to support all the pressure. Another is a muscle that encircles our gut. It looks like a cowboy went and lassoed the bottom of our large intestine and then pulled it tight. The lasso creates a kink in our bowels not so unlike a kink in a hose. When we sit on the toilet, the kink remains put, which means we have to exert more force—strain—to get the job done.

Squatting, on the other hand, unkinks the kink, giving last night's dinner a straight gastrointestinal path to speed through. Our anuses remain tension-free—barely bothered—like a pebble gently washed over by a passing stream.

"I think if you lived in India," Porrett said, "you wouldn't have any hemorrhoids."

She made it sound like people in Asia and Africa, where many still practice the squat, have archetypal puckers. (I lived in India for a year, and while I was there, I definitely perfected the squat, but I think the benefits were outweighed by the relentless dysentery.)

This is all well and good, but it poses a problem, since toilet time is sacred for many people. We go for the bowel movement but stay for the serenity. I know and respect many friends and family who would be upset at the prospect of giving up their seat for a hole in the ground—it would start to feel more like a gym class than a respite from the outside world. If my husband couldn't sit on the pot while simultaneously playing games on his iPhone, he would probably swear off the body function altogether.

The good news is that it's possible to get into the ultimate position without leaving our porcelain perch. Porrett told me that after months in a physiology lab, she and her fellow researchers figured out the riddle. If we sit on the toilet but lean forward with our elbows on our knees and elevate our feet by using something as everyday as the Yellow Pages (finally, a real reason for those antiquated telephone directories), we will be just as effective as our ancestors who took up next to a tree. There is a company that makes a stool called the Squatty Potty for exactly this purpose.

"And never push from your butt!" she said. "That's a very unhealthy way to poo."

"Then where do you suggest we push from?"

"Your abdomen," she said.

"Tighten it like you're getting punched in the gut?" I asked.

"No," she said, "take a deep breath and kind of push down."

So the takeaway to exemplary bathroom habits—and conveniently also the key to avoiding hemorrhoids—came down to two words: Don't Strain.

This essay should be over now, but a problem cropped up during the reporting. Back when I was talking to Peter Lunniss, I found out that hemorrhoids are not the worst thing to happen to an anus. I had asked him, "Is it true that anuses stretch out like the elastic on an old pair of underwear?" It was kind of a joke. I wanted it to be a joke.

But then he said, "Yes."

And I said, "Really?"

"Yes," he said.

"Dammit," I said. "That's bad news."

"Your anus becomes increasingly floppy."

"It becomes floppy?"

"Much more in women

The mindful BM

than in men," he said. "Women have a hard time compared to men with the anus."

I'd always imagined that men had the floppier anuses. I didn't realize how devoutly I'd believed that until I heard otherwise. "Can't be."

"It is," he said.

How was it possible that we are on the lower side of the wage gap *and* have a floppier butthole? If we are going to be the gender with the floppy buttholes, we should at least get paid more. It did not seem reasonable that life could be so unfair.

By "floppy," Lunniss meant that women experience more fecal incontinence, which means that the stool does not heed the sphincteric gates, but rather passes through to the other side unexpectedly.

I didn't believe this news—I refused to—so over the next week, I talked to quite a few rectal professionals. There was a lot of conflicting information, but while not everyone in the butthole community agreed, there seemed to be some consensus that women tend to be anally disadvantaged, especially if they've had babies.

This is another reason to bow down to women: Billions of them have put their anuses on the line to ensure that we have a civilization.

During childbirth, women may suffer stretching and tears on the anus. These tears are called obstetric anal sphincter injuries, which makes for the most incongruous acronym of all time: OASIS. Within a scholarly article, it sounds like this: "Whether cesarean delivery should be performed to protect against OASIS is controversial, especially as primary prevention." To wit, this is the only kind of OASIS you don't want.

These tears on the anus, which in some cases are so tiny that they aren't even visible, can sometimes have repercussions twenty to thirty years down the road. Danielle Maier, a nurse who helps run the Center for Functional GI & Motility Disorders at the University of North Carolina, said that 90 to 95 percent of her patients are female. A lot of women come into her clinic in their fifties and sixties with sphincter issues and pelvic dyssynergia. "That's the Greek term for 'My bottom don't work right,'" she explained.

"They'll say, 'But my child is thirty years old,' and I say, 'That's awesome, but after you have a baby, the damage is done,'" she said. "It's trauma plus time."

Many comedians have quipped, "Comedy is tragedy plus time." This, clearly, does not hold true for anuses.

Maier explained that the reason we don't see problems until later is that when we are young, we are able to compensate for the damage, but as we get older and weaker, the ability to compensate wanes.

If women suffer a severe tear during childbirth, the sphincters may even become alienated from one another and need to become reacquainted with outside help. Really odd things can happen to your sphincters. Things I couldn't have imagined. There's something called paradoxical contraction—the more you try to relax your sphincter, the tighter it gets. It's some of the best evidence I've seen for the argument that true evil exists in the world. For these situations, in simplest terms, you have to relearn how to poop. A technique that is often recommended is called biofeedback. This therapy helps you regain muscle control by showing you how much pressure you exert in your bottom via a screen. You can think about it as couple's therapy for your sphincters, only instead of a sofa and a therapist, there's a wire up your butt and a gastroenterologist.

"Women don't talk about this stuff," said Maier, "which is weird to me because we seem to talk about everything else."

If we did talk about it openly, I tried to envision how it would go: "Rita, I can't make it to coffee today. I've got that two p.m. anus-control class."

"Oh, that's right. I forgot. Good luck with those sphincters!"

Until that happens, know that there is a burgeoning marketplace of products—adult diapers, mostly—that help us deal with our collective crap. What was a $1.8 billion industry in 2015 is expected to grow to $2.7 billion by 2020, according to Euromonitor International, a market research firm. There is a bright side to those stats; if you're experiencing anal leakage, you can know for sure that you're not alone.

I dug for more of a silver lining, but I couldn't find one. The more I dug, the worse I felt. Women, it turns out, may be not just more prone to floppy buttholes, but also more prone to constipation. Medical researchers don't know why, yet. They suspect it's a combination of our fluctuating hormones and the fact that we are so busy taking care of other people—kids, spouses, ailing parents—that we don't put our own need to poop first. When pooping, we should follow the same instructions that flight attendants give passengers when the oxygen masks drop: Do yours first and then help those around you.

I needed someone to put all this information into perspective, so I called my aunt Karen. Not only is she a woman with an anus, she also practiced obstetrics for thirty-five years. I knew she'd cut through the bullshit and give it to me straight. You could go to her in tears about a new haircut and that woman will not indulge you; she will confirm that you need to buy a wig or go into hiding.

At sixty-eight years old, she has investigated thousands of women's crotches. When I got on the phone with her and explained all that I'd learned about sphincters, she said most of the terrible things are very rare and that I was too fixated on the extreme. "It's not like poop comes out," she said of fecal incontinence. "More often it's a brown staining."

I pondered that for a second and came to a conclusion: "I'm still not into it."

She told me that we all—men and women—get weaker as we get older. "It's a pisser," she said. "You lose hair. You shrink down. Your muscles aren't as strong, but what are you going to do? It's part and parcel of getting older."

I tried to step back and look at the bigger picture; fecal incontinence is a new milestone. If you're pooping your pants, you've made it *that* far—it's a badge of honor that signifies a long life and a well-used orifice. It's the equivalent of calluses on a construction worker's worn and weathered hands.

Nope! A floppy butthole still sounded awful.

"Look, I have a looser anus than I did ten years ago," she said. She explained that she'd had a C-section, so she definitely wouldn't chalk it up to the type of delivery.

"It's not flopping in the breeze," she said, "but sure, I notice the change and it pisses me off."

That said, she explained that it's not the most horrendous of changes. "I have a floppy butt now, too," she said of her once perky behind. She found that more disconcerting than a slightly floppy anus. "You just have to laugh at this stuff."

After hearing about the possible destruction, I realized that I had to stop worrying and start enjoying my butthole while I've still got a (mostly) functioning one. For too long, I'd been concerned that with such a finicky anus, I would have problems turning my exit into an entrance, but it was clear to me now that I needed to stop coddling the butthole and step forth to fulfill the highest of all my orifice-inspired aspirations: butt sex.

To see if my optimistic mental state matched my physical one, I once more approached Danielle Maier, the nurse at the rectal clinic at the University of North Carolina. "If you're someone who's had hemorrhoids on and off," I asked, "would you find it inadvisable to have butt sex?"

She told me that she would suggest waiting until they aren't flared, but that anal sex—if it's consensual, you use lubricant, and you are able to relax your sphincter muscles, but like *really* relax them—shouldn't be the cause of any extra problem whatsoever.

"The key is to relax," I reiterated, mostly to myself.

Then she said something that took a few days—and some counseling—to process and will no doubt stick with me until my last breath. "When people defecate," she said, "they put out stools that are larger than most of the penises I've seen, so it really shouldn't be an issue."

Yet another health professional trying to ruin things: this time the male sex organ.

Despite her comment, I remained undeterred, but before going forth, I decided that it would be fun and informative to get some pro tips from someone who has had a lot of experience with putting objects into her anus. I contacted Natasha Starr, a twenty-eight-year-old porn star who hails from Poland and has made a living out of sodomy.

On the About Me section of her webpage, you'll learn that she's tall, thin, blond, and a Libra. She has been nominated for several AVN awards—the porn industry's answer to the Oscars—all for her anal scenes. Before talking to her, I checked out some of her films (*All Things Anal, Anal Warriors, Let's Try Anal, I Wanna Buttfuck Your Daughter 16*) and was properly impressed: no orifices seemed to be harmed during the making of her movies.

Starr had a deep, raspy voice and a thick Eastern European accent. In retrospect, talking to her was like going down a double black diamond when I should have been on a bunny hill, but I didn't know that yet. "My first-time anal experience was when I was fifteen," she said, introducing herself. "My boyfriend couldn't come. I said, 'If you can go in my ass, maybe you come faster.' That's how it happened the first time."

Natasha Starr, like Wolfgang Amadeus Mozart, was obviously one of those young prodigies.

I began our interview by asking her how to get started. I'd heard that when dealing with the ass, it was smart to take baby steps. I thought she'd recommend some good introductory finger play or butt plugs.

"I like it when they put it in deep," she said, going straight to the dick, "because if you put it a little bit, it feels like you are taking a poop."

"I could see that," I said. I wanted to show her that I was on board, even though I was so off board that no board was in sight.

"On the movies they like to see the penetration," she said. "Long strokes, you know? That feels weird."

"Because then you feel like you're pooping over and over again," I said, still trying to commiserate.

"Yeah, they come out all the way and come back again," she said. "So I like it put in all the way to the end and just pound it. That is how I like it."

This was an aspect of anal sex that I had never considered.

We talked for at least another half hour and touched on subjects she found essential, like the foods that pair best with sodomy. A big steak. Yes! Lasagna. No! But overall, this was not a conversation for the everyday butt-sex dabbler. She may as well have been explaining quantum mechanics to a napping Ewok. What I found liberating, though, was how we ended our talk. As we wrapped up, I asked, "Do you ever have any issues with your anus?"

"I actually do have hemorrhoid," she said.

I was shocked. "You do?" I said. I'd seen her asshole live on my computer screen—it looked pristine, smooth and pink; it reminded me of an angel's halo.

"I do have hemorrhoid," she said again. "Maybe I push too hard? I don't know."

She was saying it like it was no big deal, like thousands of people don't depend on her to be sexy each and every day.

"Maybe it's a skin tag," I told her. I'd learned about them the previous week. An external hemorrhoid, in its wake, often leaves a souvenir behind in the form of a tiny flap of skin. Hemorrhoids, it turns out, have egos and don't want to be forgotten. The tag is nothing to worry about; it is no more than a little 'rhoid graffito communicating "I was here."

"I don't know," she said. "I think it's a hemorrhoid."

It felt so empowering to hear that not only was I not the only one with a hem, but also someone superhot shared my affliction. I thanked her for her time and honesty.

"Why?" she said. "It is not a problem. No one ever complained to me about this thing before."

I may not have learned proper butt-fuck technique—all the necessary sodomy facts I need are probably up on the Cosmo website anyway—but in the meantime, Starr demonstrated something even more important. I was impressed by how she was so open. Back when I'd spoken with Porrett, she'd said that many people are so ashamed of their butt problems—not just hemorrhoids and incontinence but also more serious symptoms—that treatable issues often become fatal because of their refusal to tell a doctor. "People literally die of embarrassment," she'd told me.

Until we get it together and put some major funding behind the development of bionic buttholes—what an exciting day that will be!—we have to do the best we can with the sphincters we were born with.

In the meantime, I'd like to think that if my grandpa were looking down on me right now, he'd have a giant smile plastered across his face. "You finally figured it out, kiddo," he'd say. "Just like with the countertops in our Palm Springs place, you've got to treat your anus with respect."

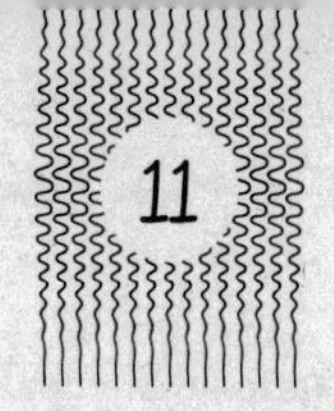

The Human Stain

I always wanted my birthmark to signify something, like maybe that I am the next vampire slayer. Some old man might walk up to me and be like, "OMG, you have the marking of the slayer on your lower back!" and I'd be like, "What are you talking about? You crazy?" At which point, he'd be like, "No, I've been looking for you since the last millennium." Then we'd embrace like old friends, followed by a cool montage of me learning martial arts, sweating at totally appropriate moments, and basically saving the world.

This has yet to happen.

The most I've gotten is, "What's that weird red mark above your ass? Looks like someone got high and tried to draw Australia."

If I'm not in fact the next vampire slayer, which is yet to be proven, I want to know what the hell is the point of having a marking. Are birthmarks just random splotches that mean absolutely nothing? How do they get there? Do they serve any purpose, and most important, when exactly is a person officially too old to start learning to be an assassin?

With the exception of symbolizing something supernatural, birthmarks seem rather inconsequential and arbitrary, which is why I didn't expect the medical community to be as hung up on them as I am. That was until I got in touch with Anna Yasmine Kirkorian, a pediatric dermatologist at Children's National Medical Center in Washington, D.C. She specializes in birthmarks, even goes around the country giving lectures on the topic.

Right away, Kirkorian threw a wrench in my belief system. "Birthmarks aren't always there when a baby is born," she said.

At the very least, I'd thought it was safe to assume that a birthmark was a mark we had at birth. In actuality, she explained, a birthmark can appear anytime from birth until the first two years of life. "What makes it a birthmark," she said, "is that it begins developing in utero."

She explained that while there are many categories of birthmark, the most common are pigmentation and vascular birthmarks. Pigmentation birthmarks occur when more pigment develops in one area of our skin. The most common types are moles, which are generally small and brown and can be slightly raised; café-au-lait spots, which are named for their coloring (coffee mixed with a large helping of milk); and the "slate-gray nevus" (previously known as a Mongolian spot), which often occurs on babies of Hispanic, Asian, and African descent. Nevi are bluish, therefore sometimes mistaken for bruises, and are commonly found on the back and buttocks.

The pigmentation birthmark was not news to me, but the vascular birthmark, well, those are an entirely different beast. They are red because they are made up of a clump of blood vessels. Blood vessels! Who knew?

The most common type is called a salmon patch, Kirkorian explained. More than 60 percent of babies are born with one of these red or pinkish blotches. "When one occurs on the back of the neck, it is called a stork bite," she said. "When occurring between the eyebrows, it is called an angel kiss."

"Wait," I said, "I think I have one of those."

Since I was born, I've had a red area on my forehead between my eyebrows. Though it's faded over the years, it still brightens when I'm

y to be popped, but instead, she asked, "Do you have lu-
n autoimmune disease that is often accompanied by bright
s.

it," I told Kirkorian, "that she was pointing to that same
n my eyebrows that we've just identified as a benign and
kiss."

aughed before releasing a short sigh. "Okay," she admitted.
marks, though lovely, can sometimes be challenging, too."
hat, I felt satisfied.

overheated or exercising. My mom always told me that it was due to my positioning in the womb, that I had my face so deeply lodged into her pelvic bone that I was permanently marked.

"No, you weren't squished," Kirkorian said. "You're just normal."

She explained that the area gets red because it's full of extra capillaries. When I get hot or angry, more blood rushes to the face and therefore the birthmark is more visible.

I was going to have to rethink my life. Who was I now that I wasn't bludgeoned by my mom's pelvis as a wee fetus? That was supposed to be my superhero origin story.

BIRTHMARK MUSEUM

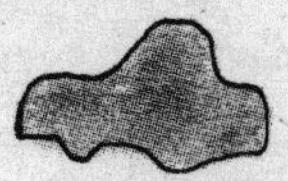

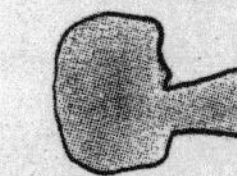

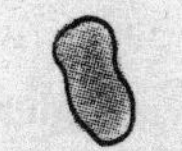

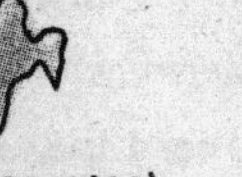

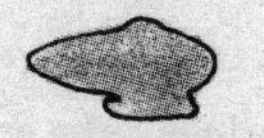

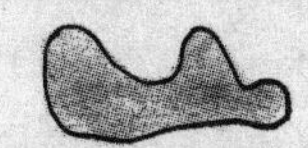

Two other common vascular birthmarks are the port-wine stain, à la Gorbachev's forehead, and the hemangioma, which, to sound less menacing, is referred to as a strawberry. It is somewhat bulbous and red like the fruit it is named after. Hemangiomas often disappear in the first few years of life, but surprise, sometimes they can grow inside the body—yes, you read that right: A birthmark can grow inside your body—and need surgical intervention so as not to disturb any organs.

I now knew some of the different types of birthmarks, but not much else.

"So how do birthmarks get there in the first place?"

Kirkorian told me that most birthmarks are caused by a mutation in our skin. While we gestate, our genes can get a little screwy, but in a totally benign way. "If you were to just biopsy that part of the skin, you would find a genetic change that is not in the rest of the body."

Research into birthmarks is still in its infancy, but it appears that they might be hereditary. "It's not like other traits where you have a fifty percent chance of inheriting it from a parent, but we are seeing that birthmarks cluster in certain families," she said. "Hopefully soon we'll figure out why that is."

This is where birthmarks get especially interesting. Birthmarks, like I'd hoped, can actually have a meaning. Now here's the bad news: If it does have a meaning, it's almost always bad. For the most part, these little skin oddities are perfectly normal, but sometimes they can signal that something went wrong during development.

"The skin is kind of a mirror on the inside of the body," Kirkorian said. She explained that only a few weeks after conception, we are a couple of balls of cells. Within those cells, there is the ectoderm, which is the layer of cells that come to form the brain and the spinal cord and also the skin. "So there is a special connection between the skin and the brain," she said. "If I see unusual birthmarks, I'm wondering if something happened genetically early on." She said that some of these birthmarks shed light on issues she might find in the brain.

Researchers have found that children with six or more café-au-lait spots are more prone to have a condition called neurofibromatosis, a disorder that causes tumors to form on nerve tissue. Port-wine stains can be

linked to Sturge-Weber syndrome
zures and sometime paralysis. Th
orders goes on and on. Kirkori
necessarily spell disaster, but it's a
"If you have something unusual li
said. "It's easy for us to take a look a

This was all interesting stuff t
was also so depressing. I'd had high h
to go in a different direction, so I t
butterflies, and slayer markings. "S
signify something good?" I asked. "Li
your lower back that looks like a ma
you're really smart?"

A short and, I admit, appropriate
Kirkorian's mouth. "That's a cute ide
plained that many cultures have a bi
some people think that moles on palm
don't know of anything scientific that's

I think she was saying that I might
my own.

"Look," she said, "I think birthmar
right there." She went on to pontificate
"They are gorgeous in their many uniq
adding individuality and diversity amo

Kirkorian continued to go balls-out
body markings—"They are decorative, li
but as she did, I interrupted to tell her the
in my life.

I told her about my friend Kat, who c
at the middle of my forehead. I thought
that she liked my new bang trim or, at th

whitehead rea
pus?" Lupus is
red facial rash
"It turns
redness betwe
beautiful ange
Kirkorian
"It's true. Birt
And with

The Eleventh Toe

It's early fall and I'm in the fitting room at Macy's Herald Square location, trying on some sleek pants—black, snug, and made of some ubiquitous stretchy material. Those puppies fit my gams like maple syrup atop a short stack. I spank my own behind and then spin in circles while chanting "cha-ching." I just won the pants jackpot! But then when I turn back to face myself in the mirror, something happens. My eyeline draws toward my groin.

My vulva has vacuumed up the teensiest bit of fabric, creating the shape known notoriously as camel toe.

I suspect that I'm the only one who would notice this deviant crevice. We're each our own worst critic, after all, and it's not like it's the Grand Canyon. I step outside for a second opinion. Without pointing toward anything specific, I ask the salesgirl what she thinks.

"I love those pants," she says, but then she purses her lips to the side while raising one eyebrow, "but I don't know that they *work*."

A week later, I'm walking up Sixth Avenue with a friend when we spot a blond drenched in spandex; a conspicuous wedgie has crept up her front. I ask my friend what she thinks of the situation.

"I judge," she says. "I always judge." She tells me that not even an attractive woman can pull off being attractive if she has a camel toe. "First, I assess what I consider to be the damage, like on a Richter scale for earthquakes. Is it just a hint of camel toe? Is it full-on up the vag so you can see two perfectly formed vag mountains? Is there underwear involved, and if so, which kind? Basically, I'm trying to figure out how this happened."

The camel toe we saw, according to her, was a 5.8.

Two weeks later, I'm standing on a corner in the East Village while my dog sniffs the behind of another woman's dog for so long that it's getting awkward—we humans need to at least acknowledge each other. One thing leads to another thing leads to me finding out that this human is the CEO and inventor of a new underwear brand called Camel No. It is a type of panty featuring a thin silicone insert that obfuscates the outline of the vulva.

"It's not like it's only for women with a loose, gaping vagina," Maggie Han, the statuesque dog owner and outspoken proponent of seamless crotches tells me in a tone that makes me think twice about the possible gaping-ness of my own vagina. "We are all prone to toe, and toe is sacred," she tells me. She's so intimate with the phenomenon that she's dropped the "camel" part of the term altogether.

I am intrigued. I get her contact info.

When I get home, I look up Han's brand and find that it isn't the only camel toe–prevention panty out there; there's also Camelflage.

In my life, I hadn't given toe much thought. I kept it simple: I knew it was something to avoid. But after those consecutive run-ins with the infamous cleft, I began to think more deeply about the phenomenon. Inconsistencies abounded. Toe—the contour of our sacred parts—is exactly what you'd expect to see under skintight clothing, so why are we all repulsed and surprised when it happens? I'll go out in the city with my ass cheeks dangling just below my cutoff shorts, but the mere outline of twat is a fashion no-no? It doesn't make much sense. I was feeling very Seinfeld about the whole thing: I mean, what's the deal with camel toe?

To find out why people had such strong opinions about labial skin and how it should be tucked (or untucked) below the belt, I began sleuthing in earnest. I expected the fashion crowd to know why toe was so abhorred, and I reached out to Hollywood stylist and TV personality Emily Loftiss, for whom the mere utterance of the two words caused discomfort. "It makes me instantly cringe," she said. "Can we move on to the next subject now?"

"What's so bad about it?" I pressed.

Loftiss wasn't able to elucidate. "It's just no no no absolutely not ever no," she said.

I contacted athleisure-wear designers. I wanted to know if there were tricks they used to prevent camel toes when designing leggings. But no one wanted their brand linked in print with vulvar cleavage. Most didn't even respond, but Lululemon public relations coordinator Adrienne Watson at least wrote back: "We would like to respectfully decline at this time." Another international brand would agree to speak only on background. That is the same request people make of journalists when they fear political retaliation, or jail time, or even assassination for their remarks.

The fashion world, in other words, completely shut me down.

I moved on to the average woman—maybe the everyday jegging-wearer would have thoughts on our aversion?—but the ladies I spoke to, like me, thought of toe solely in the context of how best to avert it. There was the thirty-five-year-old who told me that she'll wear a panty liner like it's a force field. Then there was Jessica, a thirty-four-year-old Broadway dancer, who thought she was more prone than others. "My theory is I get it because I have a ba-donk-a-donk," she said, referring to her large ass. She suspects that her pants, in order to cover her substantial behind, have to make a shortcut through her front side. "So I always buy a bigger size than I want," she said. Another chronic toe sufferer, twenty-eight-year-old Angela, has a unique method to counteract the vulvar infringement. "I shimmy the majora up a little bit and then I pull the labia minora out a bit," she said, of her outer and inner lady lips.

"What does that do?" I asked.

"I'm evening out the landscape, know what I mean?" Essentially, she explained, she builds a wall of pussy each morning so that her pant seam can't trespass her labial gates. "It never works for very long," she admitted.

Tactically, this was all very interesting—I'd never thought of such innovative fixes—but I was still hoping to peel away another layer of the toe onion, like why we all consider it so important to prevent.

I needed to find someone who could talk about toe with nuance, someone who had spent years weighing its philosophical and sociocultural importance. Yes, it was time to call up Han of Camel No. She was clearly an expert on the matter. After all, she's made it her career goal to foil toe. She must, therefore, know what's at stake when our pants deceive us.

We meet at Boulton and Watt, a hip restaurant in Manhattan. (Tabletops are made from distressed wood. There is exposed brick. Quinoa is on the menu.) When I spot her, she stands up and twirls around. "I wore some real toe creators so you could see how my underwear work," she says. Her black pants are so tight they may as well be taped on, yet she's correct; when I investigate, I do not see a trace of genitals.

She tugs them upward from the waist. Still nothing. I am impressed.

Over a beer, we chart the phenomenon (and backdrop for her burgeoning business) back to its origins. One hundred years ago, there was only one meaning for "camel toe"; it referred to the actual toe of a one- or two-humped animal known to live in arid regions. But then two societal shifts occurred, which gave the term its new meaning. The first was that we women began wearing pants, some of which were so tight that the contour of our anatomy became visible. Second, the Brazilian wax became popular. "The bush used to deflect seams from going up," Han said. "Now there is no buffer."

I'd always had the feeling we weren't giving the bush its due respect. Besides its ability to double as a suds-generating loofah, there are other pubic hair benefits we don't talk about; when I get completely waxed, my pee stream is unwieldy, like a broken sprinkler. The bush, I realized, acts as a route, enabling streamlined and tidy urination. I didn't bring that up, though, because it didn't seem possible that this five-foot-eleven woman with Barbie contours and hair that looked like polished obsidian would be able to relate to a body function gone haywire.

"But why a camel? It's like the ugliest animal," Han mused. "Couldn't it be named something cuter like 'koala paw'?"

"Fair point," I said.

"Society likes to give us a hard time," she said.

I asked if certain crotches were more prone to the toe, but she said in her experience that's a huge misconception; among all vulva owners, toe is an equal-opportunity invader.

"It has to do with ill-fitting clothes," she told me, "not someone gross and loose." Mothers call her all the time and try to qualify why they need an order of her underwear. "I'm like, 'Listen, lady, you don't need to explain away about your kids and how loose it made you. I have no kids, I'm just tall and not stretched from a baby, but I struggle with the same issue.'" It was then that Han slapped the table to relay her sense of shock. "I can't believe the extent some women go to get rid of toe." She told me that some get labiaplasty—plastic surgery for their vulva wherein bits and pieces of the inner or outer lips (or both) are snipped off. I'd known of the surgery, but it seemed too drastic to believe that toe might be the sole impetus for someone to go under the knife. "No need to anesthetize anyone," Han said. "Let's just do a Dr. Scholl's here."

Han, I discovered, didn't just do this for a quick buck; she approached her invention as if it might have as much impact on gender equality as Gloria Steinem did. She said that her underwear empowers women to take their mind off what they look like and focus on what they're doing. "I want women to be free, I don't want them to be in the middle of a kickboxing class and be like, 'Fuck, dude, there's my toe!'"

But as we talked more, I felt disconcerted. I appreciated that Han was solving a problem and giving women solace where they have concern, but I also couldn't help feeling disgruntled. Not only do we have to put our breasts in bras to protect innocent bystanders from a rogue hard nipple, but soon we'll all be expected to own specific gear to keep our vag lips on lockdown as well.

"What if women feel like it's just one more thing they have to do?" I said. "Do you feel like you're playing into this at all?"

In other words, in twenty years, will women feel stifled by their caged vulvas and, in revolt, burn their Camel Nos? It seemed that by hiding that little crevice indefinitely—making it forever invisible—we might lose something else, too, but I couldn't put my finger on what exactly.

"You can feel however you want about it," she said, "but I think it gives women peace of mind." To her, walking around with camel toe is lewd and inappropriate. "There are kids outside"—she pointed to the window behind her. "And men can get sucked in. It's like a tractor beam."

"Does that mean that you think toe is sexy?" I asked.

I'd always considered toe to be something goofy—something that passersby chuckled about rather than something that they stared at forlornly. Toe, to me, doesn't have the allure of cleavage—maybe because it's unintentional. The woman touting her cleavage wears it deliberately, or at the least, with awareness. The woman with camel toe does not. She has the equivalent of the inadvertent plumber's crack. She becomes fair game for a recurring thread on Perez Hilton's blog, Guess the Celebrity Camel Toe! In one such picture, Iggy Azalea is shown wearing tight orange spandex shorts. "Oops" is written in white and an arrow points to her two little lumps.

But Han saw it differently. For her, it is something sacred that should be guarded and kept private at all costs. Only the most intimate of relations should know its whereabouts. "That's your boyfriend's toe. That's your husband's toe. It was your father's toe," she said, bringing home her point. "Toe is too personal for just anyone."

"I think my husband would laugh if I had toe."

"He might laugh, but then he'd say, 'Let me get at that toe!'"

We had to agree to disagree. To Han, toe was too provocative, something that bystanders shouldn't see. Our crotch tremors—anywhere from 2.8 to 9.3—were an entirely personal affair best left inside our underwear. To me, toe was awkward and embarrassing, similar in feeling to smacking into a very clean piece of glass that I'd thought was an open door. But what really struck me about this conversation and what I continued thinking about as I walked back home was the women Han had mentioned earlier, those women who so loathed their toes that they got surgery. If that is true, that is some serious dedication. I had to know, what was actually happening up on those stirrups and inside those doctor's offices?

I cast my net wide, contacting seven plastic surgeons who were known for trimming and "rejuvenating" women's genitals. I contacted so many because I figured that if I was lucky, maybe I'd get one to call me back. I thought that by the nature of what they do, they would be guarded. They were mutilating women's vaginas—ruining our individuality by making our genitals match some porn-star prototype. I judged the surgeons, and so did my friends. We got up on our high horses. When I found that altering women's genitals was the fastest-growing surgical aesthetic procedure in the United States, we said to one another, "Can you believe this bullshit?" But to my surprise, all the surgeons I asked to interview were more than happy to give me time. They were generous and open. It was all very confusing. They didn't think they were doing anything wrong.

After I chatted with four surgeons, I decided to visit one in particular, Dr. Red Alinsod, a pioneer pussy stylist. I wanted to learn why he does what he does as well as experience a vulva consultation for myself—how would he propose getting rid of my camel toe?

I arrived at his headquarters in Laguna Beach, California, at seven a.m. on a Thursday. He's so busy and booked with dissatisfied vulva owners that he had to meet me before his office officially opened. Before the consultation, we chatted at his desk. There was nothing in the room that gave away his profession. I had been hoping there'd be something worth noting, like a bronzed statue of his favorite vaginal creation with a placard that said "Best in Show," but instead there were only papers, books, and a computer. Alinsod wore blue scrubs and had an affable countenance, and he always bookended his lighthearted comments with "That's a joke I like to tell." He also had such a sincere face—cherubic cheeks and a near-constant grin—that, and I never thought I'd say this about the arrangement of anyone's facial features, I would immediately trust him with a scalpel near my clitoris.

Alinsod wasn't always in this line of work. He's a urogynecologist by training and used to do primarily reconstructive surgeries—"When a vagina prolapses, it looks like an inside-out sock is hanging between your legs, and my job was to put that back where it belongs"—but now his business is made up of 90 percent elective aesthetic vulvovaginal surgeries. "It's like art," he said. "I get to create something that's beautiful and functional after the trauma of childbirth or even changes of menopause, or whatever the reason is."

Because he enjoys the surgeries so much, he chose to work in a beach community. "You have to provide what the area needs, and Southern California in particular is very, what do you call, focused on appearance."

I was surprised he said that; it was just so unbelievably honest. He was pretty much saying he targeted a population inclined toward self-consciousness.

Alinsod is known especially well for his "Barbie Look" procedure, which includes cutting down the labia minora so that the little wattles don't peek out from the majora, which are the puffy outer lips that grow hair. The medical terminology for the procedure is *labiaplasty*, and it is the most popular of all the vulvovaginal surgeries, of which there are about six variations.

"So you call it the Barbie surgery?" I asked. When I saw the term on his website, I was annoyed, almost offended.

"The terminology came about when I was practicing in Los Angeles, before I moved here," he said. "I didn't invent it. My patients would just come into my office and tell me, 'I want to look like Barbie.'"

"But Barbie doesn't have a vagina."

"That's my exact point right there," Alinsod said. "She does not have a vagina or labia, so I would say to the patients, 'You want to look like Barbie?' And they would say, 'Yeah, I want to look just like Barbie,' and I would say to them, 'Barbie doesn't have labia,' and they would say to me, 'Exactly.'"

He explained that all the surgeries he performs were borne from his patient's desires. In other words, he simply supplies the demand. "I had women asking me, 'Can you do this? Can you do this?'" he said of the days before labiaplasty became more mainstream.

When I did more research, I suspected more and more that women want this surgery because of what they see in porns and pop culture.

Mature minoras are often airbrushed out of pictures, giving the idea that models do not have them, but it's not just aesthetics; some say it's also practical because long minoras can become uncomfortable—can get twisted and chafed—during exercise and even sex.

After we talked a bit more, I said, "Are you ready to take a look at mine now?"

On the way to the consultation room, we passed many paintings, which were all abstract interpretations of the female body. The pieces seemed to divulge a subconscious message: We, too, are artists who can reshape the human form.

I changed into a blue smock and sat on the examination table until Alinsod returned with a nurse. He flipped up the steel stirrups. I placed a foot into each. "Scoot all the way down to the very end," he said, as he handed me a small mirror so that I could follow along with an up-close view of my genitals.

Earlier, he'd told me that Brazilian waxes were so popular that he hadn't seen pubic hair in decades. In fact, he blamed waxing for the uptick in vulvar surgeries, which was something I'd heard from other surgeons. "Shaving started this whole business," he said. "We didn't focus down there the same way when it was covered with hair, but now people can see that they have longer minoras and asymmetry, a wide clitoral hood, or a mons pubis that is puffy."

So not only is the lack of bush causing more severe camel toes, as Han had noted earlier, by no longer acting as a buffer between pant and twat crack, but it's also supposedly some of the impetus behind some women's desire to hack off what's revealed beneath. In other words, seeing their genitals without any hair in the way has caused some to be more critical about, and therefore also want to change, its appearance.

We need to save ourselves. Bring back the bush!

As Alinsod leaned forward to check me out, I felt like I had to warn him. "I know it's been a while," I said, "are you ready to see some pubes?"

"You're completely fine," he said, laughing.

I laughed, too, but then he ruined the mood when he said, "I actually see hair all the time on my elderly patients."

With the end of a Q-tip, he traced my anatomy, showing me all the parts and what he can do to them with a scalpel. Though all the procedures pertain to different areas of the vulva, there was a common theme: to reduce, to shrink, and to lessen. The penis-clad among us have the exact opposite desire: to enlarge, to enlarge, and (wait for it) to enlarge. It made me think about my subway commute: Men manspread onto multiple seats while women take up as little space as possible.

After a while, he moved down to the bottom of my genitals. "That's the perineal skin," he told me, pointing to the area between my vaginal opening and anus. The skin was thin and had folds so small that I wouldn't have noticed them had he not pointed them out. "It's a normal skin fold," he said, but he mentioned that many patients have him remove the minuscule wrinkles. "They think it looks like a hemorrhoid and they don't want that hemorrhoid look."

Ridiculous! I knew what real hemorrhoids looked like, and these little folds were kid's play. These folds were for novices. Clearly, some people should not be allowed access to hand mirrors.

Alinsod went over several more procedures—the minimizing of the clitoral hood and vaginal tightening—but then finally got to the topic that had initially inspired the visit.

"So if I said, 'I don't like my camel toe,' what would you do for me?"

He took one side of my outer lip between his gloved fingers. "As you get to your fifties, this starts to sag like this," he said, pulling my lip away from my body. "And the space between here and here becomes longer and longer," he said, pointing at the base of my vulva to the apex of the stretched lip.

He went on to explain that to get rid of camel toe, he makes an incision along the inside base of the lip and another along the ridge of the lip where the hair starts to grow. "I remove that whole section," he said, pointing to the inside of my outer lip, the part that is pinkish, a bit shiny and sensitive. "That's a lot of tissue," he said.

He told me that about 5 percent of his patients are born with an extra fat pad inside the outer lip. "It makes it look like their majoras are testicles," he said. "It's huge. They get huge and make women feel very self-conscious."

"So would the surgery be worth it for me?" I asked.

He knew this was a journalistic endeavor and I was speaking hypothetically. "I would tell you that what you have is a normal variant and you absolutely do not need surgery," he explained. That being said, he told me that he'd be happy to operate on me if I wanted it anyway. "If you wanted to reduce it for cosmetic reasons, just like people with completely normally functioning noses want to remove a hump on the bridge—you have to think of this that way—I could help with that."

He told me that he does an average of four to six camel toe–ectomies each month and that none of the surgeries, as long as they are performed correctly, should affect sensation during sex. "The women are so pleased afterward."

I attempted to verify Alinsod's claim, so a few days later, I spoke with a woman, thirty-three-year-old Nicole Sanders, who had had labiaplasty of the minora and majora as well as a clitoral-hood reduction. Another labiaplasty specialist, Michael Goodman, had hooked me up with her. Besides feeling comfortable wearing tighter pants, she said she's also a lot happier. "It used to cause me pain and insecurities," she said about what she considered to be her oversize vulva. "I didn't want anyone to ever look at me when I was naked, but now that has changed. I feel really, really proud. It may sound very strange to say that, but I feel proud about how I look now—really confident."

I still felt so skeptical, but if these vaginoplasties made women more confident, who was I to say that it was wrong. Maybe the surgeries weren't as bad as I thought?

I visited Dr. Lara Devgan at her office on the Upper East Side of Manhattan. I specifically sought her out, as she's a self-identified feminist. In other words, she's not just another dude who cuts pussies off. She's a woman who believes fiercely in women's equality but who also performs roughly eight to twenty vulvovaginal procedures a month at seven to eleven thousand dollars a pop. I wanted to understand how she rationalized performing them.

Upon my arrival, Devgan and I retreated to her inner sanctum, a quiet space away from the bustle of her practice. The green walls were dotted with tasteful light fixtures, giving off the flattering glow that is customary in high-end fashion boutiques. This woman was clearly a powerhouse; she managed to have a successful plastic surgery practice while being a mother of four, and she was currently baking her fifth beneath her white doctor's coat.

As I sat across from her, I was trying to balance my desire for deep and skeptical inquiry with my awe—she had clearly mastered life.

After a half hour of small talk, I finally hit her straight on: "How is this something that is right and good for women?"

She went forth into a staggeringly persuasive rant. "Labiaplasty is one of those things that if it's not bothering you, it sounds crazy," she said. "It sounds like female genital mutilation and it sounds like a third-world phenomenon and sounds like it's something that should never happen, but if it bothers you and you're looking in the mirror every day and you're trying to decide how you're possibly going to wear that swimsuit for your high school swim meet or how you're going to be intimate with your partner on your wedding night or how you're just going to get through your day and get on the subway without adjusting yourself in public and without causing bleeding or chafing in your private parts. If you're in that camp, this is not something that seems frivolous. But you know, those people who have the luxury of thinking that this is something that is flippant and unnecessary are people who happen to have the luxury of not being in this situation."

After mulling over what she said, I felt guilty and closed-minded. I had arrived at Devgan's office hyperjudgmental of others and now I was leaving hypercritical of myself. Maybe I'd been able to view the world only through my own very specific vulvar experience. Who was I to judge, anyway? Maybe I was part of the problem. Women are shamed enough for not reaching impossible physical standards; should they then be shamed again if they try hard to attain them?

Alinsod, when I'd posed the same question in his office, had another effective argument. He'd told me that he used to constantly get calls from angry feminists, telling him he was the harbinger of evil. "It was really weird to me because here are feminists calling me to get me to reduce choice for women," he said. "If I would have listened to them, I would have eliminated a chance for women to improve their lives. It doesn't make sense to me."

He told me about a twelve-year-old girl who was on her school swim team. One day, when she got out of the pool at a meet, her inner labia were sticking out and down the side of her leg for everyone to see. "Her mother was horrified," he said. "She ran to cover her with a towel." The mother, soon after, rushed her daughter in to see Alinsod—to remove her inner lips—in order to save her daughter from future humiliation.

Even though I was pleased that the young girl didn't have to be embarrassed any longer, I also felt something more complicated well up. There was a disconnect. She was embarrassed not because there was anything wrong with her, but because of how we've come to narrowly define femininity—that our genitals are the right size only if they happen to fit into the itty-bitty crotch space allotted inside an itty-bitty bathing suit. Something about that felt uncomfortable and reductive to me.

Our vulvas are somehow supposed to defy the laws of physics. They, with their many folds and abundant flesh, are expected to maintain the same look as a doll made of seamless plastic. It's like asking an English muffin to forgo its quintessential Nooks & Crannies. The Nooks & Crannies are what makes the English muffin an English muffin.

So after speaking to surgeons, people who profit from what they do, I realized that I might be getting a biased opinion. It was like asking an oil company about the impact that dredging has on the environment—"All the animals are totally fine! Don't worry!" Metaphorically speaking, I needed to get the perspective of an environmentalist. Were the animals really okay?

I trolled books and studies for reliable pussy conservationists and quickly connected with Jane Caputi, a women's studies professor for more than twenty years and the author of *Goddesses and Monsters: Women, Myth, Power, and Popular Culture*. I called her up to speak about camel toes and whether or not labiaplasty was leading us down a dangerous road.

She, in turn, dove right into a short history lesson. "One of the words for vulva is 'pudenda,' and 'pudenda' comes from the Latin root of which word?"

I knew the word—had even used it before—but was unsure of its origins. "I don't know," I said.

"'Shame,'" she said.

I didn't believe this—it seemed too impossibly "on the nose"—so I looked it up on my computer as we spoke, and surprise, surprise, this woman who had written four books on feminism was correct.

"It comes from shame," she reiterated. "Women are supposed to be ashamed of their vulvas." She said that in the distant past, many cultures believed in the goddess and that the vulva was a symbol for power; it was considered sacred. "Not sacred as in untouchable and pure," she said, setting me straight. "We're talking the Force." She went on to explain that patriarchal cultures took the vulva's sacredness and made it obscene and repulsive. "One could very much imagine a culture where showing the outline of the vulva could be considered very erotic and attractive and clothing would be designed to highlight that, right?" she said.

I could see it. I mean, why not? That had been one of the things that confused me from the start. We think breast cleavage is hot, so why does hot suddenly stop at pussy bulge?

"But see, the vulva epitomizes female power and that power is seen as threatening to patriarchal control," Caputi said.

In essence, she was telling me that my own repulsion of camel toe went way beyond what I thought was simply a fashion faux pas; my disapproval of it on myself and other women has roots—gnarly, ugly ones—in our culture. When we snicker at a woman's vulva outline, we are actually in some weird, fucked-up way condemning and constraining our own gender.

"Women have been taught or told to associate their vulva with vulnerability, silencing, and moments of powerlessness," she said.

"So what do we do now?" I asked.

"What the tactic has often been is to take what they shame and realize they are shaming it because it's a source of power and beauty—whether it be skin color or hair or a vulva—and reclaim it as something powerful and beautiful."

"So are you saying we should all go around with camel toes?"

"Well, first off, I hate that name," she said. "Let's say instead 'showing off the luscious labia' or 'the power of the labia.'" She told me that sporting the look would be brave and awesome, but even changing the terminology around it would be a good start.

I visualized a scenario where I run into a woman's eleventh toe on the street and then I attempt, in my mind, to see how I'd put Caputi's new lexicon to work: "Excuse me, your wonderfully luscious labia are sticking out of your pants."

Something told me that while headed in the right direction, I didn't get it quite right.

Another fairy godmother of female genitalia is Rebecca Chalker, who is an outspoken critic of vaginal surgery. It was easy to find her: Just Google "anti labiaplasty" and go down the black hole. She's the author of *The Clitoral Truth: The Secret World at Your Fingertips,* a book that more or

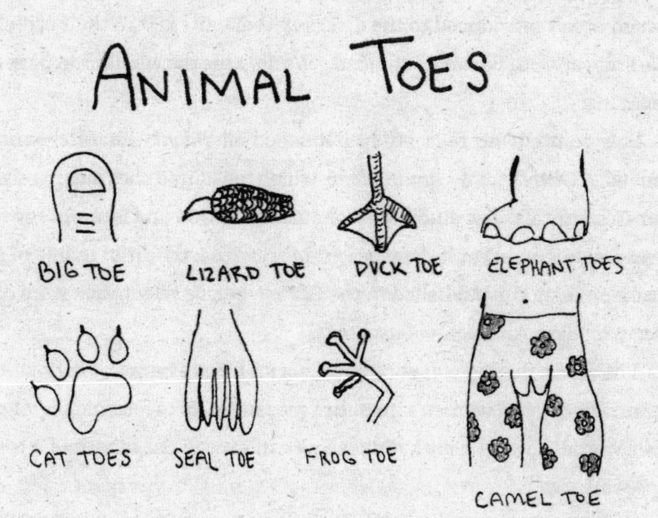

less makes the case that the clitoris should win the Golden Globe for best piece of human anatomy. When we began speaking, I still had complicated feelings about the surgery—it seemed drastic and unnecessary, yet I appreciated that it helped many women feel better about themselves. Seeing that I was torn, I asked Chalker a question straight out of Alinsod's playbook.

"How can you say that this surgery is bad when it's giving women a choice?"

Chalker, it turns out, doesn't believe that there is an actual choice. She thinks we have only the illusion of options. "Say I'm offering you a choice between two things: apples and oranges," she said. "So if I go into the next room and I say, 'Take your choice,' but there is only a bowl of apples, what do you say?"

"I say, 'Where's my oranges?'"

"That's the flaw," she said. "That is the flaw of that argument. It's not a choice! They aren't offering women a choice; the choice is that their genitals are freaky." Chalker went on to explain that the main problem is that women aren't introduced to the diversity that can exist in their crotches. "We grow up with Barbies!" she said. "We only see the versions in porn and magazines."

She pointed me to a study published in *BJOG: An International Journal of Obstetrics & Gynaecology*, which measured the range in size of women's genitals. The study was conducted in 2005 and is one of the first to research female genital variation. Meanwhile, the first article of this nature on men was published in the 1800s—before telephones were common in modern American households.

The study showed a vast range of normal. For example, in the sample size of fifty vulvas, women's inner lips measured up to four inches long. In other words, normal healthy inner lips can exceed the length of a toilet-paper roll.

"But there's no honoring the diversity of women's genitals," Chalker lamented.

Both women, Caputi and Chalker, mentioned the importance of not blaming or judging women for getting the surgery, but rather blaming our culture, which gives us little space to be who we are.

Caputi said it best: "I would never be against someone making a choice to do something that makes them feel empowered or feel good about themselves, but I would rather work toward a world where women are not so systematically shamed about their bodies, where we don't have such fear and loathing in particular of female sexuality and anything including large labia that smack of female power."

I'm usually not all about the patriarchy keeping us down, but I surprised myself with this one; I was basically on the phone with one hand up in the air as if she were delivering a sermon.

"Having choice isn't enough," she continued. "You also need access, access to a safe and healthy culture that tells you, 'You are okay, Mother Nature made you, and you are beautiful.'"

A few days later, I received an email in my inbox. A man by the name of Nick Karras explained that he'd heard that I was investigating vulvas. Apparently, the world of pussy preservationists was quite small and word had spread like herpes in a freshman dorm. "You should definitely come over and check out the Petals Project," he wrote. He said he had thirteen years' worth of research to share.

My interest was piqued. The following week, I showed up at Karras's home. The sixty-five-year-old greeted me at the door wearing beige cargo shorts, a blue plaid button-down, and bare feet. He escorted me to a brightly lit back room, which was sparsely furnished; only two high-back chairs faced the wall. On that wall, there were twenty-four framed photographs, each a tasteful close-up of a vulva in sepia tones.

Karras reached his hand out toward them. "That's a portal to the other side," he said. "That is where the souls come through. It's just so powerful. It's the core of femininity. It's everything. That's why there's a lot of issues around it, because it's such a powerful part of being a woman."

We sat in the chairs and stared at the folds of skin that make up our most private part. In total, Karras has taken 268 photos, and he assured me that each was different from the last. "There can't be a beauty standard, do you see?" he said. "Each one is entirely unique."

There were chubby ones, tiny ones, wavy ones, dark and light ones. One in particular caught my eye: the labia like a big undulating scarf tumbled freely between two legs.

Though he's been called a pervert by art institutions and blocked from showing his work in gallery openings, some academics were enlightened enough to use Karras's photographs in a study. Dutch researchers found that college-age women who were exposed to these photographs had higher self-image scores than the women who were not. Seeing the diversity inherent in each of us made them feel better about themselves.

Karras and I spoke for almost two hours. Our conversation went all over the place—from his past relationships, to nudist colonies, to how butterflies choose their mates—but the thing that really stuck with me was his opinion on meaty vaginas. What women are paying thousands of dollars to change, he celebrates by calling them well hung or endowed, the terminology usually attributed to men. "You know, I personally like larger," he said. "There is more to play with."

"And that's a good thing?"

"Oh my God, that's over-the-top cool," he said. He put his hand on his chin, as if he were in Athens giving a philosophical treatise down by the agora. "Not that I'm not saying that a small petite one isn't fun also," he said, "but I notice that it's really cool if the woman really likes it and is flaunting it—it's like a man who's flaunting, like a man swinging it around. He likes his big nuts. And women who have the big labes and really like them, oh my God, there is something to that."

Not many days later, while I was walking up the street to get some tacos for lunch, I spotted a homeless man walking toward me, but just before passing on my left, he stopped and pointed at my stretch pants. I was really into these pants; a smattering of bold and vibrant spring flowers crawled up and down my legs upon a background of crushed brown fuzzy fabric. Some might call the pair flamboyant. Since this guy was kind of stylish himself in his bright green albeit grease-stained and holey polo, I suspected he was about to compliment me on the loveliness and colorfulness of my getup. Just as I was saying, "Thank you," accepting what I saw as his inevitable compliment, he instead lifted his left leg and pointed at his own pants this time. "Camel," he said. He cleared his throat before continuing. "Have you heard of camel toe?" I must have been putting camel toe vibes out into the universe, because never had the toe been such a large slice of my life's pie chart. It was almost like when you've gone thirty-five years without hearing the word *haberdashery* and then as soon as you do, it seems like the word is everywhere, even your five-year-old nephew is singing it over and over again to the tune of "Row, Row, Row Your Boat." Except that I'd heard of camel toe before and instead of my nephew singing about it I was standing on a street corner engaged with a homeless man who was judging the way my pants formed to my lower lips.

As I took in the situation, the dude giggled, put his leg down, and then quietly continued on his way.

First I looked toward the man ambling off into the distance, and then down at my own crotch. I didn't see anything too overt, but spotted what may have been construed as the two digits of an ungulate. For a moment, I contemplated taking action—some sort of excavation work—but instead I shrugged my shoulders and continued up the street.

You're So Vein

My day was not going particularly well. It was a Friday afternoon in 2009 and my landlady at the time, a seventy-year-old with a white curly fro who wore nothing but ankle-length nighties, stood just outside the window of my Greenpoint, Brooklyn, apartment, belting my name in her thick Polish accent. "Mara! Mara!"

She had upended my garbage onto the sidewalk and was busy classifying each scrap. "This," she yelled, holding up a milk carton, "not recycling. No!"

This tiny woman with outsize energy had escaped communist-era Poland and now put all that vigor and passion for survival into making sure I didn't screw up trash collection. Suddenly, she was on her hands and knees, inching toward a floss segment. "What is this?" she said. "Cannot put here. Cannot!"

The commotion caused a small crowd of furry-faced and flannel-bedecked hipsters to circle and eye my refuse. One tub of empty hummus would have been okay, but nine was making me reflect on my life choices.

Also, I couldn't focus on my work; each time I would settle back at my desk, she'd yell, "Mara!" again. One such time, I found her holding up a plastic tampon wrapper—apparently those can't be recycled, either. I was just trying to help the planet, but she was looking at me as if I were the second coming of Stalin.

For a day like that—a crap one with large doses of humiliation and negative levels of productivity—I'd developed a ritual to turn things around: blood donation.

At that point, I had been donating blood for ten years, since I was seventeen. No matter what was going on in my life—a breakup, a failed exam, a fiery-hot razor burn—it always made me feel better. It managed to retroactively trivialize my own troubles, because it put everything into perspective. It was a way to focus on the big picture. Suddenly, I'd be dealing in the real stuff, the life-and-death stuff.

Plus, most people, to save lives, have to go to medical school for eons and take out exorbitant student loans, but I just lay on a cushioned platform for fifteen minutes and ate Oreos. I'd feel accomplished even though I had done absolutely nothing. It was a lazy way to be a really good person.

When I felt low, giving blood also gave me a little ego boost. I would exit the bloodmobile with a bright red bandage wrapped around my arm. Friends and random people like the bodega clerk would eye the bandage and ask, "What happened? Are you okay?"

I'd shrug like it was no big deal, but I'd feel a tiny flash of moral superiority. "I'm fine. I just gave blood, that's all."

If you're careful and don't take a shower, the bandage will stay on for up to two days.

On this day, I found out that the bloodmobile—a standard RV retrofitted into a medical clinic—was parked on the corner of Bedford and North Seventh in Williamsburg. After the twenty-minute walk from my apartment, its loud, humming generator muted the car horns and cacophonous conversations from the pedestrians just outside.

Once I signed in, I sped through the usual rigmarole—the questionnaire, the short interview, and the hemoglobin check. Everything was still going according to plan. My angst would be minimized and resolved in no time.

Next, a phlebotomist had me lie down on a pleather bed. Some people don't like to look, but I'm a watcher; I always inspect every aspect of the draw. She wrapped a large rubber band tightly around my biceps and then placed a foam ball into my hand. She told me to squeeze it as she located a vein. "Ah, you've got a great one right here," she said, fingering the underside of my elbow.

I took it as a compliment—I always do—even though I'm pretty sure that nurses use that line on everyone.

The nurse punctured my arm and then taped the needle down while my blood made its way through a translucent tube and into a plastic pouch tacked to the side of the bed.

There really were so many things to love about giving blood. While helping humanity, I was also on a mini diet that worked in a matter of minutes. A pint of blood, after all, is at least one pound. It must be acknowledged here that my stance is not widely accepted: "Giving blood should not be thought of as a weight loss tool," wrote Carol Ochs, a health reporter with the Associated Press.

After ten minutes, the pouch reached capacity. The nurse unearthed the needle and bandaged the wound, and then I stood up for my favorite part of the donation experience—the free and unlimited cookies. But when I reached the snack box, something was amiss. My brain had embarked on a solo sojourn. I became light-headed and dizzy. Voices became muffled, and a dark, fuzzy cloud fogged my eyes. I'll never forget how everything disappeared—I could no longer see the bags of trail mix I'd reached out to grab. I raised my hand, a deeply ingrained impulse apparently, to ask a question about my state, but it was too late.

A minute later, I woke up facedown in the narrow aisle surrounded by a puddle of urine and two irritated phlebotomists. I had passed out and peed myself on the floor of my once beloved bloodmobile.

That is the only time in my life I can recall regretting sweatpants as an outfit choice. In that context, their main quality changed from "comfortable" to "absorbent."

In my opinion, the nurses should have called a rescue helicopter and a trauma therapist, but instead they asked me to rest for only a few minutes before sending me on my way. I must not have been thinking properly, because I showed up unannounced at Dave's apartment. At that point, Dave and I had been dating for a very short time. I didn't even know that he loathed celery yet, but there I was at his door in my light gray sweatpants, a color, it turns out, that makes pee stains look as conspicuous as a hickey planted on a pale neck. He looked me up and down. He did not break up with me. He even let me inside. It was one of those harrowing moments in life. I'm pretty sure it all happened in slow motion.

I passed the night with a deep and abiding migraine.

Fifty-six days later, which was the first day I was eligible to donate again, I got a call from the blood bank to set up an appointment. Once you donate, the blood bankers are on your veins like bill collectors are on a late payment. Usually, I'd be delighted. They needed me! No one else needed me. I didn't have a pet or a kid. I worked as a freelancer. The majority of my days consisted of waiting for emails. Sometimes, I'd wait so long for an email that I'd send an email to myself just to be sure that my email was still functioning. The email, without fail, would land in my inbox a second after I pressed Send. Sad story, but true: Each time, for a moment, I would forget that I'd just sent an email to myself and actually get excited at the *ding*.

But this time, after I'd fainted, I wasn't in the least bit flattered when they called. I shouted, "No!" and hung up the phone. During the next several months, these blood bankers sweet-talked me. "But you have O positive and we really need O positive," one said, trying to seduce me.

But donating blood, I'd decided, was pure evil.

Over the next several years, I had bad days—a killed article, a bad review, a fight with a friend—but I wouldn't dare give blood again. To feel better, I resorted to other healing activities like retail therapy and emotional eating.

It's now been seven years since that wretched day. In the interim, Dave and I got married—which should count as a win to women everywhere who have shown up for a date drenched in their own urine—but I still haven't stepped foot into a blood bank again.

I've stayed away because fainting is, for lack of a more sophisticated adjective, scary. One minute I was chilling like a boss up on my high horse, about to chow on some cookies, and the next minute I was on the floor being stared at with pity by strangers. My brain shut down mid-thought like a phone dropping a call. I lost control of my body, even the ability to control one of my most basic functions.

My reaction to fainting isn't unprecedented. In fact, no one I've met thus far has enjoyed spontaneously powering off. I was in the sauna with a friend when he fainted. First his eyes rolled backward and then his body crumpled to the floor. When he awoke a few moments later—still lying on the tile and gazing up toward the ceiling—the first thing he said was, "Did everyone see my balls?"

Fainting makes us vulnerable. While those around you are present, for you, those moments of unconsciousness will be forever unaccounted for. After two years, that same friend is still fixated on those few seconds. Whenever I see him, he always asks me the same exact question: "But really, did you see my balls?"

I have another friend who fainted during a meditation course. Not knowing exactly what brought it on, she was concerned she would go unconscious again, maybe this time while she was driving. She didn't man a vehicle for months. She still gets panicky thinking about how her brain went black. Yet another woman I spoke to fainted while being screamed at by her abusive boyfriend. While he berated her, her body slackened and fell to the floor. Still not knowing why she fainted, she asked me recently, "Do you think my mind was trying to save me from experiencing that awfulness?"

I, like my friend, had also theorized why I lost consciousness—I thought the phlebotomist had taken out too much blood and so my body could no longer operate. Separately, I theorized that maybe my body went into shock and needed a time-out to recuperate. I've come back to revisit this moment, because it occurred to me I'd never done any research. Everything I believed about fainting, well, I'd totally made it all up.

When I looked at the situation with some distance, I saw how crazy it is: We are all walking around with a spontaneous off button hidden somewhere inside and yet have no idea what triggers it. Maybe if I figured that out, I could finally regain my confidence, go back, and start enjoying one of my favorite pastimes.

Over the next week, I was able to speak with two fainting specialists. One was Dr. Nicholas Tullo, a cardiac electrophysiologist who runs the New Jersey Center for Fainting. The other was Raffaello Furlan, a cardiologist from Milan, who coedited a book on the phenomenon.

Fainting, formally, is known as syncope (*sin*-kuh-pee). Upon first hearing the term, I felt misled. It sounded like a quaint fishing town in Alaska rather than what it is: falling onto your face after a loss of consciousness.

But by talking to both researchers, I would learn that syncope is not always evil and is in fact quite common. Half of all Americans faint at least once during their lifetimes. Fainting is also simpler yet simultaneously more complex than I'd expected.

"It's really due to one thing," Tullo said.

He explained that while there are many different triggers—emotional as well as physical—that can cause fainting, each results in the same condition: a lack of blood flow to the brain. "Without blood, your brain cells just shut down. It only takes four to ten seconds to completely lose consciousness."

Without blood in my head, it made a lot of sense that I couldn't eat cookies. The surprising part was to hear that I could actually survive those moments without going brain dead.

"There is no danger," Tullo assured me. He went on to say that while fainting is the result of a problem—the lack of blood to the brain—it also results in the cure. When we become horizontal by falling to the floor, our brains fill back up with blood, because the fluid no longer has to fight against gravity to get there. (Of course, this is the case only if you're not driving at the time or near the edge of a steep cliff.)

"You recover almost immediately," Tullo said. "Crazy enough, a lot of people even wake up feeling refreshed."

I know nothing of refreshment; those who feel good after slamming to the ground must be of the same breed that cheer with stupefaction each time their airplane touches down.

But syncope became much more complicated when Tullo began to explain what causes that lack of blood to the brain in the first place—the impetus for the off button to be pushed, so to speak. It turns out that researchers still don't completely understand why it happens.

Furlan, the other expert, told me that the answer is elusive. "It's like trying to hold tiny grains of sand in your hands." And he's been studying syncope since Jimmy Carter was president.

What researchers do know is that our brains control many involuntary reflexes, such as breathing, digesting, temperature regulation, and blood pressure. Reflex syncope (which also goes by the name vasovagal syncope), the most common type of fainting, occurs when the brain sends out a wonky signal to lower the blood pressure and/or heartbeat. When the blood pressure drops dramatically, blood can no longer make it all the way up to our brains—hence the loss of consciousness.

"But why these reflexes go wrong," Tullo said, "we really don't know yet."

So after donating blood, it wasn't that I didn't have enough blood left in my body or that I was in shock, it was that the blood I did have wasn't distributed properly. Something, though it's hard to say what, exactly, had triggered my blood pressure to rapidly drop.

"What do you think happened?" I asked Tullo after detailing my experience.

"A lot can predispose someone to a faint," he said tentatively. He listed off a few of the many possible triggers—dehydration, standing too long, a hot day, or a claustrophobic environment.

It turns out that there are so many things, both emotional and physical, that can cause a faint. There is even something called defecation syncope, which is the term for fainting if it happens while taking a crap—in this case, the brain, misreading a message sent from your colon, triggers this precarious blood-pressure drop.

"But you know, often it has to be a perfect storm," Tullo said, "like that old song goes, 'When the moon is in the seventh house and Jupiter aligns with Mars.'"

"Fainting is like Aquarius?" I said, surprised that any musical, let alone *Hair*, would figure into this.

"Yeah," he said. "See, you're dehydrated and you're standing up and it's hot and someone gives you some bad news or you have some emotional stress—everything lines up, and that's when you'll faint."

He told me that doing the same activity on a different day or under different circumstances would rarely lead to the same outcome. "You can go ahead and give blood another time and it won't happen," he said.

He told me that, statistically speaking, I also shouldn't be too concerned about fainting again. After a fainting episode, one has a 40 percent chance that she will faint again in the next year, but if she doesn't faint, then her chances drop to 17 percent for the year after that. This is because, he said, fainting often occurs in "clusters," but as to why that is, we still don't know. "Some people never faint again, but some do so several times a month," he explained. "It's one of those imponderables." Considering that as of the time of this writing it's been about 2,555 days since my one and only faint, this news was heartening.

Before I was satisfied, though, I had another question. If fainting was so common, I wondered if it had any benefits. I thought about my friend, the one who'd fainted while in a fight with an ex-boyfriend. She'd suspected that her body shut down to save her mind from having to undergo and catalogue that experience.

"So does fainting protect us in some way?" I asked. After all, reflex syncope is not considered a disease, but something that actually falls within the bounds of normal human behavior.

Tullo told me that theoretically one could say fainting is protective, but they would be using poetic license, because no benefits have been proven. "In my opinion, the brain is not trying to save anything." He explained that fainting occurs because the brain is screwing up. "By signaling the blood pressure to drop, it's responding inappropriately," he said.

Furlan, who has looked at the phenomenon from a more teleological standpoint, had a different take, but even he was hesitant to attribute too much purpose to a fainting episode. "It's not protective in this era," Furlan said, "but maybe it was to our very old ancestors."

He explained that when some animals are frightened, they freeze. "Rabbits in the desert respond to a rattlesnake by playing dead," Furlan said. "A snake doesn't like to eat dead meat, so maybe the rabbit will survive its predator."

When an animal "plays dead," it is actually experiencing an involuntary rapid downshift in heartbeat and blood pressure. It's possible, Furlan said, that when a person faints because of emotional triggers—fear, shock, disgust, phobias, and gruesome images—they are actually experiencing a relic of the playing-dead response found in our fellow vertebrates.

How helpful fainting may have been for our ancestors would have depended on the kind of predators they ran across. "Am I going to respect you because you are lying down and not dangerous, or will I take the opportunity to finish you and cut off your head?" Furlan said, positing the decision to be made by a potential predator. Ultimately, he said, figuring out if there were benefits when a person fainted in the Paleolithic period is impossible. "Who can know?"

In essence, it's unlikely that my friend's brain was trying to save her from experiencing the quarrel, but it is possible that she felt threatened enough that some ancient mechanism was trying to save her from being eaten.

All I knew for sure at that moment was that I felt like a jerk. I didn't realize how standard it was for people to faint because of emotions. I'd always suspected that people who fainted at the sight of needles or upon hearing shocking information had a low threshold for suffering. More or less, I considered them weaklings.

"So those people aren't just extra wussy?" I asked Tullo.

"No, it's not their fault," he said.

I still didn't believe him. "But it sounds really precious, you know?"

"But it's not," he reiterated.

Apparently, fainting from emotional stimuli is often genetic. Some people can't stop fainting at the sight of a needle any more than someone with allergies can stop from sneezing during a hay storm. I'm going to try to have more compassion next time I see a movie where the ingénue faints at the sight of a sharp-toothed and slimy alien.

Before Tullo and I finished up, I asked him one more thing.

"So how do you trust your body again?"

He was quiet for a moment. "I know it's scary," he said, "it's always scary to lose control of your body, but know that you'll always wake up again." He went on to say that though the triggers are still somewhat mysterious, there are actions to take, which can make fainting less likely. The best preventatives are drinking lots of water and eating salt. While staying hydrated helps increase blood volume, salt helps to retain the fluid, making a drop in blood pressure less likely. "People think salt is bad," said Tullo, "but it's only bad if you have high blood pressure."

Regular exercise is also important, as is listening to your body. "If you start to feel the prodrome," he said, of the sensations like dizziness, blurred vision, muffled sound, sweating, and nausea that tend to precede fainting, "lie down immediately and put your feet up."

Just because you feel faint, it does not mean fainting is inevitable.

If I had known that before, instead of wasting time by raising my hand to ask a question, I would have hit the ground intentionally.

"Again, all you need to do is get blood back to your brain," he said.

Furlan was also supportive of my return to the donation chair: "Go give blood. It won't happen again, I assure you."

But he had slightly different advice, which was not to think about passing out. "When you're there, just look out the window and say to yourself, 'Isn't it so pretty outside of my window?'"

A week later, I booked an afternoon appointment at a bloodmobile.

The hardest part of the appointment was picking out my outfit. I believed those fainting fellas, but I wanted something comfortable that would—just in case they were wrong—also work well with urine. After trying on various ensembles, I went with an ankle-length black skirt and a white T-shirt long enough to cover my groin.

When I arrived at the bloodmobile, Donna, a petite middle-aged woman in a blue nursing uniform, brought me into a small room for the vetting process. Even though it was so cold that my arm hairs stood up like a million tiny pins, my armpits had managed to produce sweat stains the size of dinner plates.

After passing the interview and hemoglobin test, I warned Donna about what had happened the last time I donated. She didn't seem fazed. "Some people, when they black out, can even have a bowel movement," she said.

I guess that's life; people are always trying to one-up you.

Instead of hightailing it out of there, I thought back to something a bit odd yet comforting that Furlan had told me. He suspected that those who were crucified back in the day actually died from fainting—without being able to move their legs, blood would pool into the lower half of their bodies, triggering syncope, but because they were pinned to a cross, it made it impossible for them to fall and get blood back to their brains again. I think it was his quirky way of making me feel better: Unless the phlebotomist crucified me afterward, which was unlikely, I would survive even the worst fainting episode.

Donna decided to take some extra precautions. She had me lie down and put cold wet towels around my neck. In order to keep me conscious, she exploited part of my thermoregulatory mechanisms—when a cold compress is used on the body, blood vessels will constrict in that area of the skin. Making my neck cold, she explained, would free up more blood to circulate in necessary areas like the brain.

From there, the experience was predictable. Phlebotomists, though it had been almost a decade since I last donated, were still complimenting people's veins. "You've got a nice big one right here," the woman working on me said.

She pressed the needle inward. I looked out the window. A few plastic bags were being swept upward by the wind, but I followed directions anyway: "Isn't it so pretty outside of my window?" I said to silent and skeptical stares.

After ten minutes, the phlebotomist extracted the needle. She'd successfully obtained one pint, which I soon learned accounted for more than a tenth of my blood. That seemed a bit excessive, but when I later contacted the American Red Cross, Ross Herron, a divisional chief medical officer, explained that the standard donation is one pint, or more specifically 525 milliliters (15 percent of the blood volume of a person who meets the minimum weight requirement of 110 pounds to donate blood), because that is how much blood a person can lose without having too many physiological changes. "In other words," he said, "it is a safe level of blood loss." Once someone loses more than 20 percent of her blood volume, all hell begins to break loose—the heart rate increases as the heart pumps harder and still will not be able to distribute blood to all the necessary tissues. "This is when a blood transfusion becomes essential for trauma patients," he explained.

My takeaway: Make a concerted effort not to get run over by a car after donating blood. Your buffer has already been depleted (though don't worry too much; blood volume does quickly increase with rehydration).

After resting for a few minutes, I stood up, wondering if my vasovagal reflex had been triggered. I was still standing after a minute, so that seemed positive.

The phlebotomists still didn't trust that I had it together, so one followed me when I told her I had to go to the restroom. She wouldn't let me lock the door. "If I hear a thud, I'm coming in!" she said. You would think that having someone follow me to the toilet would feel invasive, but I actually liked it; I felt special yet also in danger, like a high-value target. It made me think I wouldn't mind being the president.

I stood up. I wiped. I flushed the toilet. When I washed my hands, I was still standing upright! Donna, at that point, felt that I was in the clear. Before going home, I walked to the front of the bloodmobile and did what I'd aimed to do so many years before: sat down and indulged in three bags of mini Oreos.

The cookies were good, but I've got to be honest: After all that buildup, staying conscious and not soiling myself felt incredibly anticlimactic. When I got home, even Dave was slightly disappointed. I never realized how much I'd been terrified by the incident, but also oddly cherished the memory. After all, showing up at Dave's door in that state had been an essential step in our love story. It was a yarn we'd shared multiple times with close friends and family. All I can say is I guess sometimes the moments in life that feel the worst in real time can make for some of the best ones when they are assembled in the highlight reel.

Within an hour, I got over the fact that I'd had a very routine and unexceptional experience. I was also pleased that I didn't have to do laundry.

Over the next several weeks, I busied myself with work. Thoughts of blood donation didn't come up, at least not until yesterday, when I received an email from the blood bank: "Your blood donation was sent to Gadsden Regional Medical Center in Gadsden, AL to help a patient in need."

I thought of a woman lying in a bed with a needle in her arm, exactly how I had been lying in a bed with a needle in mine, but instead of blood shooting upward through the translucent tube, that very same liquid—liquid she needed to live—would flow downward and into her body.

It was for her that I'd be going back again.

Bloody Hell

A couple of months ago, Dave and I were standing in the kitchen, deciding what to make for dinner. As we discussed baking salmon, our discussion somehow took a sharp turn.

"Why haven't you gotten your driver's license yet?" I said.

Dave is forty-two years old. Driving is not that hard.

We moved on and began to consider making a stir-fry, until I was like, "And you still don't put the toilet seat down!"

At that point, we decided it was probably best to order take-out. Dave pulled out his laptop and began scrolling through the possibilities, but I cut in while we were debating the type of cuisine.

"Why haven't you gone to the gym lately?" I asked. "You're going to give yourself a heart attack."

I'm a chubby-chaser, and because of that, I live in a perpetual catch-22: My preferred body type could also cause harm and shorten the life span of its owner. I experience a near constant and conflicting mix of wanting to feed my husband a tub of lard and wanting him to go run a marathon.

"You've had that gym membership for months now," I added.

Dave is conflict averse, so when I bring up subjects with volatile properties, he tends to ambiguously nod or pivot the conversation. "Chinese, Mediterranean, Thai," he read from the screen, not engaging with any of my complaints.

"I can't have a fight by myself," I said. "I need you to state your position."

"Vietnamese?" he said. "You like pho."

I was so completely not in the mood for pho at that moment, yet he was still looking at me like it was possible that somehow I might want pho. How, after four years of marriage, could he look at *this* face contorted in *this* way and think that it meant "Yes, I want pho"? After a few more empty and looming seconds, I said, "You don't even know me!"

He sat there perplexed, his hand steady on the mouse. He looked wounded and slightly distressed. "I don't understand," he said after a moment. "Why are you shitting into my soul?"

I didn't have a good answer—shitting into his soul felt so right at that moment. I had a knot in my chest—a tightly wound spool of yarn, of words—and something inside me had green-lit its unraveling.

But when he called me out—*why are you shitting into my soul?*—I also began to see myself from the outside. I felt like there was another me who was observing myself through a window. I could see my mouth like a little volcano, erupting with angry words, and I started to regret my intensity. I loved the dude, brows furrowed, in front of me. I wanted to rewind, to put that spool of yarn back where it came from and make everything okay.

At times like those, I pull out my calendar to see when my next period is due to start. Then I ponder how many days out—what is my window exactly?—that I'm allowed to blame my morose and erratic mood on the infamous yet possibly mythological premenstrual syndrome.

Dave put in an order for Szechuan tofu, kung pao chicken, and soup dumplings. As he flipped on the television, I tapped him on the shoulder.

"Maybe it's PMS," I finally said. I think that may have been my attempt at an apology.

I don't know much about PMS—I'm not even sure it actually exists—yet I use those three letters liberally. I use them as a scapegoat for a range of homicidal impulses, gray-cloud thoughts, and potentially divorce-inducing statements. Those three letters also give me hope: If it's my PMS making me say awful things, feel awful things, and do awful things, then it's not actually *me* who's at fault for any of it, so clearly everything will be better and back to normal in the morning. Or the next morning. Or the next next morning. Or the next next next morning. I mean, right?

Given the role PMS seemed to play in my life, looking into it felt dangerous. I know many women, myself included, who depend on PMS to rationalize a crappy day, to justify a meltdown, and very importantly, to account for why she bawled at the end of a completely predictable movie with horrible acting and no artistic merits or originality whatsoever.

So, what if I found out that PMS wasn't real and that horrible movie actually moved me? When I told a friend I was going to research the psychological aspects of PMS, she got pissed.

"Don't take that away from me," she said. "I'll fuck you up. It's very real for me, dude. It's a great handicap."

I had to know, though, when I claimed that PMS made me do it, what was I saying exactly?

For a few days, I had a one-track mind. I asked every woman I ran across her views on PMS. I got as many answers as there were women I asked.

"Do you get PMS?" I asked my friend Sara.

She cocked her head to the side and said that she wasn't sure. "You should ask Nate," she said, referring me to her partner.

My cousin Nora said she totally gets PMS. "I cannot be trusted during that time," she said. "I turn into a raging lunatic." She told me that during one particularly intense episode, she left her boyfriend, who had been annoying her, at a gas station and drove away.

My friend Reyna had an optimistic twist on the monthly episode. "I loathe PMS," she said, explaining that it makes her feel like a two-ton ball of shit that should put itself out of its own misery, "but I love that it gives me an excuse to eat hamburgers." It's only when her womb is about to have its bloodbaby that she feels like she finally deserves to satisfy a craving.

When I asked another friend, Mariah, she responded, "Isn't PMS kind of like lactose intolerance and being gluten-free?"

"How do you mean?"

"Like once you hear about it," she said, "you suddenly seem to get it."

"I don't know." I shrugged. "Maybe."

PMS seemed to have as many rumors swirling around it as the popular girl in high school does about the state of her virginity.

Once I began trying to pin down the facts about PMS, all the hearsay made a lot more sense. Even among scholars, the topic is incredibly divisive. There are as many studies proving that it exists as there are studies proving that it does not. Also, a hunch I'd had turned out to be true: The exact cause of PMS is still not known.

Kimberly Yonkers, a professor of psychiatry at Yale and the director of the Center for Wellbeing of Women and Mothers, has been researching PMS for a quarter of a century and thinks we are closer than ever to an answer, but prefaced what she said with, "Opinions vary." She explained that the current theory is that the fluctuation of hormones—more specifically, metabolites of progesterone—cause some premenstrual women to experience withdrawal. Hormones then do not cause the mood swing; rather, the mood swing is caused by each woman's differing sensitivity to the monthly hormonal shift. "We think some women are biologically more vulnerable to experience more severe symptoms," she said.

There is also a premenstrual state called premenstrual dysphoric disorder (PMDD), which is more or less the supervillain version of PMS. PMDD is what happens if your PMS fell into a radioactive spill and transformed into the Joker. Each month, PMDD launches its victims into a temporary but deeply depressive abyss.

Though there is no blood test to confirm the presence of PMS, the diagnosis is calculated clinically just like depression or bipolar disorder.

"So I can't just feel shitty and say it's PMS?" I asked Yonkers.

"That would not be consistent with PMS."

"Even if the shittiness I feel is right before my period?" I asked.

"Not if you haven't charted your symptoms for several months to see if there is a pattern."

To have legit PMS, it turns out, you've got to feel like crap in a very specific way. Yonkers said that you have to show a symptom or symptoms for most cycles in the year, which she considers seven out of the twelve months, and the symptom or symptoms have to be bad enough to cause a negative impact on one's life. "A functional impairment of some sort," she explained. "Fights with a significant other or trouble at work or school."

The symptoms also have to fall within the right time period: They have to occur no more than two weeks before you start to bleed and subside no more than a few days after your flow begins. The timing piece of the mood swing is often misunderstood. If a woman is acting "bitchy," an a-hole might say, "She must be on the rag." In this case, he would be attributing her mood to the wrong issue, because PMS—namely the "premenstrual" part of "PMS"—means "before menstruation." This person, if he's going to say anything (which he shouldn't), should actually be saying, "She must be pre-rag."

Yonkers explained that only 10 to 15 percent of the population would qualify for actual clinical PMS.

PMS, by this definition, seemed too constraining for the breadth of my moods.

Soon after talking to Yonkers, I asked Dave what he thought—did he see a pattern in me?

He answered very tentatively, as if it were a trap. "There may be times when my soul is shat upon that seem to coincide with where you say you are in your cycle," he said, "but I am not one to correlate the two."

Though PMS is an actual diagnosis, it is clear that many of us have adopted the term and, for better or worse, use it more liberally.

All the factors that added up to PMS and the diagnosis itself seemed so arbitrary. I wanted to get more clarity on the issue and figured that if I went back in time and found its origins, just like I did with sweat and vaginal odor, I would be able to better understand its significance for women today.

First I contacted Helen King, a professor of classical studies at the Open University in the UK, who specializes in ancient Greek gynecology.

"When did people first start to recognize PMS?" I asked.

King said that ancient Greeks were aware of physical changes that occurred before menstruation like breast tenderness, but they did not recognize any mood changes. "Greeks thought women were crazy, full stop," she said. "So getting your period wouldn't make you any more crazy."

"Of course," I said, "women were crazy all of the time."

"Exactly"—she laughed—"you know, those out-of-control women!"

Then she went on to tell me things that don't necessarily have to do with PMS, but that I have to relay to you anyway, because the information was so insane that it deserves a slight tangent.

King explained that Greeks in the fourth century BC thought that menstruation was essential to female health. "If you didn't release the blood, you would just explode," she said, "not literally explode, but really, you could die of suppressed menstruation." Because of this, ancient Greeks had various methods to make women whose periods were late bleed. King shared two of her favorite recipes, which could easily double as torture techniques. "In the first, poisonous beetles were used," King explained. Once the head, feet, and wings—the most toxic parts—were removed, a half a dozen of said beetles were wrapped in wool and then inserted up the vaginal canal. The idea was that the woman—beetles up her cooch—would experience such irritation that she would scratch at herself until she bled. "Then you could say, 'See, she had her period!'"

Upon hearing that, I could muster only a mere "Jesus."

"I know!" she said.

Beetles were used when the blood was stuck, but if a doctor also suspected that the uterus was floating around the body (because that's what uteruses did back then—they floated around the body wreaking havoc like a wayward hobo searching for his next pit stop), they'd use a special technique to lure the womb back into its right spot. Once it was in its rightful location, it could then bleed.

The technique: puppy fumigation.

First, a puppy was disemboweled and stuffed with aromatics and spices. Then it was put into a jar and buried in a hole that had been prepared with hot charcoal. Finally, the rim of the jar was plastered with clay, but not before passing a straw through it and into the vagina.

"Then you sit with the vapors from the cooking puppy passing into your womb," she said, "and this supposedly inflates your womb like a balloon and then it floats around in your body until it gets back to where it should be."

I was full of hope when I asked this next question: "When you say 'puppy,' did that word mean something different back then?" I thought perhaps it could be a colloquial term for a throw pillow.

"I'm afraid it means 'very small dog,'" King said. "That's what it means."

She made a really big deal out of the fact that women had authority over the straw. "They could put it as deep as they wanted," she said. "Women didn't usually have any control in ancient medicine."

That's how bad it was for women back then: Getting to decide how far up your vagina to put a straw that was spitting dead-puppy fumes was considered to be quite progressive.

After hearing about these methods, I was not surprised that ancient Grecian women were considered crazy—I'd be agitated, too, if a doctor, at any moment, could shove toxic insects up my genitals.

Clearly, some horrific practices existed in ancient Greece, but interestingly enough, there was no mention of PMS, so I jumped forward about a millennium to see if I could find something there. As with vaginal odor historians, there aren't a ton of PMS historians, either, but it just so happens that Michael Stolberg, the medical historian at the University of Würzburg in Germany, whom I'd spoken to earlier about sweat's sketchy past, had also spent a year of his life poring through letters, medical records, and journals from the fifteenth to nineteenth centuries in search of the origin of premenstrual suffering.

He found that depending on the notion of menstruation of the era—remember, until relatively recently we didn't know menstruation comes from the uterus shedding its lining each month—women's premenstrual symptoms and complaints shifted accordingly. Then Stolberg offered another complicated lens through which to glimpse PMS. "Like any disease," he said, "people give meaning to the symptoms, and those meanings can be personal, they can be cultural, and they change with time."

In the sixteenth century, the period—or monthly flux, as it was called—was considered to be women's garbage-disposal system. Throughout the month, women collected bodily impurities and stored them in her uterus until, all in one go, she would flush the toxins through her crotch. Men didn't have to worry about their toxins, because conveniently for them, they weren't considered inherently dirty.

Our premenstrual symptoms at that time—cramps, colic, and something mysterious called "strangulation from the uterus"—were seen as validation that what we were passing was highly noxious. Think something akin to the gunk stuck to the sole of a shoe after it's walked through Chernobyl's nuclear sludge.

In the late sixteenth century, Stolberg told me, the idea evolved. Physicians came to believe that women suffered from an overabundance of blood. The period was a natural way of expelling the excess—it was like letting air out of an overly taut balloon. "In that time," Stolberg said, "women felt tension all over their bodies and a sense of being overfilled."

Headaches, for example, were considered a sign that excess blood was being collected and brought downward from the head.

In a wonderful twist, men could also suffer from similar premenstrual symptoms, because they, too, could suffer from excess blood. Because they didn't have such a convenient orifice for the blood to exit, they had to depend on erratic nosebleeds and aggravated hemorrhoids. (Who wishes they had a vagina now?) "They, as well as women, would only feel relief upon the onset of bleeding," explained Stolberg.

There were slight permutations on this theory until the eighteenth century, when a drastic shift occurred: The uterus itself, not the blood, became the monster. When irritated by blood, the organ would go on a rampage. The way Stolberg described it made me think of an enraged rodeo bull trying to buck off its rider. "It was an entity in its own right," he said of the uterus, "and it could get very pissed off."

"So let me get this right," I said. "The uterus got angry and then took it out on the woman?"

He said it was actually more an irritation than a full-on fistfight: "The uterus got extremely irritable and irritated the whole nervous system." It reminded me of poison ivy, the way the plant can irritate our skin, but in this scenario, the uterus was the irritant and agitated the inside of our bodies.

The premenstrual symptoms a woman experienced, which for the first time regularly began to include erratic moods, were evidence that her uterus was having a tantrum. The woman would respond to her upset uterus by becoming hysterical.

A modern version, such as in my case, I suppose would look more like a woman yelling at her husband to put the toilet seat down or tearing up at car commercials—the part where the dad hands his daughter a new set of keys and watches her wistfully drive off without him for the first time. When I cry at that part, I won't even check my calendar. I just have to assume that my hormones are wacky. It's the only way I can live with myself.

After talking to Stolberg, I realized that as recently as the nineteenth century there was still no sign of the three letters—P, M, S—that we now associate with heating pads, unchecked chocolate consumption, and bitchiness. With a little more research, I discovered that by looking as far back as ancient Greece, I'd gone back way too far. My timing was so far off that it would have made just as much sense to take a time machine to the Ice Age in the hopes of killing Hitler. PMS seemed so deeply rooted in our culture that I had suspected its origin was ancient—I could easily envision Cleopatra lopping off five heads and then saying, "Sorry, y'all. PMS!"—but in reality, the term wasn't coined until 1953 by a gynecologist named Katharina Dalton. PMS, if she were a person, would be just old enough to enjoy her senior discount at the movies.

Though PMS, as a name, stuck, it wasn't the first time premenstrual symptoms were classified. Only a couple of decades before, in 1931, a man named Robert Frank noticed that some women—not all—lost their shit before their period. They were "reckless" and engaged in "foolish and ill considered actions." He called this condition "premenstrual tension," or PMT. He did some rather extreme stuff, stuff that makes puppy fumigation look kind of cute, to try to heal these women from their crazies. He thought that by decreasing their estrogen he could restore their sanity, so he tried radiating their ovaries, extracting their ovaries, and also stuffing their bodies full of male hormones.

In a later paper, a scientist would critique the latter tactic: "It occasionally happens that the dose needed to relieve symptoms is high enough to cause acne, the growth of a slight moustache, and lowering of the voice." Menstruation, the paper also noted, often stopped.

Being one who already has a slight mustache problem, I got some serious chills.

PMT didn't catch on very widely; mostly, it seems, because it suffered from a case of bad timing. As Karen Houppert reported in her book *The Curse: Confronting the Last Unmentionable Taboo: Menstruation*, in the 1940s the military began recruiting women to aid in the World War II effort. She wrote that because of this, the whole idea around menstruation shifted. Women suddenly looked up to Rosie the Riveter with her biceps flexed. They were told that they were strong, agile, and dexterous—their periods and premenstrual pains couldn't stop them from doing anything. Even informational videos were made to teach women that premenstrual suffering was nothing more than folkloric balderdash.

In one video, called *Strictly Personal*, the doctor who narrated tried to set them straight. "Some twentieth-century girls still believe that lavender-and-old-lace hokum about no activity and no bathing during menstruation," he informed female recruits. "That's Victorian stuff. And so is that trash about nerves and sensibility during this period."

As soon as the war ended, though, and men streamed home from the front, Houppert explained, women were pressured to give up their jobs and take up their place back at home with the kids. This was aided, she wrote, by a new slew of studies claiming that "the workplace was potentially hazardous to women's unborn children, and that women's cycles made them less-competent workers than men." This occurred around the same time that Dalton came on the scene with her theories about PMS.

At this point in my research, I had to take a short break. I should no longer be surprised that history had so many unsavory things hidden within it, but I was. I didn't like the idea that we were allocated menstrual issues depending on the convenience of society and the job market at the time. PMS was all starting to sound rather manipulative and fucked-up.

The more I read, the more this Dalton lady seemed like a real piece of work. She conducted studies and published papers stating that premenstrual women could temporarily become shoplifters, accident-prone, suicidal, a danger to their children, and violent, as well as be the cause of divorce.

In her book *Once a Month*, she wrote that women's abilities degraded staggeringly during "that time of the month." Women, therefore, "can be assigned to less-skilled jobs such as packing and stacking during their vulnerable days," she advised, "rather than remaining on tasks that are much more complex." She also wrote poppycock like this about the premenstrual woman: "She may buy totally inappropriate dresses that don't fit and are the wrong color, and which she will never wear."

Nonsense, I say! Any woman who has ever set foot in a clothing store knows that no matter the day, buying dresses that are deemed ugly once they are brought home is an inescapable part of the shopping experience.

In the 1980s, Dalton made headlines again when she stood as an expert witness on two different murder cases. She asserted that because the two women were premenstrual when they slaughtered their victims—a coworker and a lover—they should be given less time, which they were indeed granted. While she tried to explain that only the most extreme instances of PMS caused violent psychosis, the criminal cases and the resulting media blitz nevertheless managed to popularize the term "PMS" around the world.

My panties, if I hadn't been commando when I was reading about Dalton, would have been in a gigantic uncomfortable wad. Basically, she came up with some real cray-cray stuff. If she presented her theories today, I would imagine she'd be met with rotten eggs and hundreds of angry op-eds, but as it is, whether we are aware of it or not, she formed the foundation for what we now think of PMS.

I wish life were straightforward, but unfortunately, it's not. See, Dalton, just as easily and depending on perspective, could be viewed as a hero. Women who had complained of premenstrual symptoms—bad moods and episodes of anger—felt like they had been dismissed by doctors and told that what they were feeling was "all in their heads." Dalton was one of the first who validated women's feelings by telling them that what they were feeling was real.

The thing is, though, she then went a step further: She determined that PMS was a medical condition that could be fixed. After Dalton, premenstrual symptoms were no longer something that were considered normal and expected before a period; they had become pathologized—something to be treated.

When I was born, my mom had a hysterectomy, so by the time I heard her PMS stories, they had already turned into lore. Now that I was taking a closer look at PMS and had learned some of its questionable origins, I wanted to talk to my parents again to better understand their take on the cycle. Surely, their outlook in some way had rubbed off on me.

I was sitting on my sofa in New York City, looking down over a crowded street—taxis honking and a glut of meandering pedestrians. Meanwhile, my parents were at their home in North County San Diego, where they've managed, for the past forty years, to foster and revel in one of the best relationships I've ever witnessed. They are so inseparable that I call them Keena—shorthand for Ken and Deena.

After warming them up on the phone with some easygoing chatter, I asked them how they'd dealt with PMS back when my mom had ovaries. Their honesty was surprising if not a tiny bit disconcerting.

"As I recall, we interacted very poorly because I would be a total shit all of a sudden," said my dad. "I'd be doing everything wrong. I'd be coarse, callous, and offensive."

"Sounds about right," said my mom.

"After a certain amount of time, it was clear that it would happen monthly," he said. "She would hate to have it be the thing, you know, because this is her—this is how she feels."

"He has it pretty good, doesn't he?" my mom said.

"She'd say, 'How could it be a chemical thing? It isn't and you're a shit to say so,'" he explained. "I think it took us a few years to get onto this Midol thing."

Even though Midol has no psychotropic properties, my parents swore that the medication was the cure to a bitchy woman. Literally, it is just Tylenol and caffeine. Maybe the medication relieved cramping pain that made my mom feel irritable, or it's possible she experienced the placebo effect. All I know is, when I was a teenager, they'd both tried to push it on me as well.

"So you would tell Mom to take Midol when she started acting cunty?"

"Yes, and then she'd take my head off," he said, "but then after a while she started to allow that it did help her and helped us get along better." After a pause, he added with a slight laugh, "It seemed to improve my behavior considerably."

I could just imagine my mom on the other side of the country, feigning annoyance by knocking him on the shoulder and playfully rolling her eyes.

"So, Mom, how did you take that when Dad was like, 'Go take a Midol'?" I asked.

"I really felt demeaned when he would say that," she said, "but I felt like I was not in control of myself. I was acting unlike my normal self and maybe I was flaring out at people and I should take something to help that out. Sometimes I didn't want to take the Midol. I wanted to voice my—"

"She wanted to express her womanly power," said the male half of Keena.

"Why do you call it that?" I asked.

"She was expressing herself," he said. "'Guys are shits and they try to take the power.' She was getting her power."

I had to imagine that my dad's interpretation—womanly power—had something to do with the era in which this took place, the 1960s and 1970s, when there was a big feminist push for equality. A few decades earlier, when a woman's foremost role was to be a mother, the renowned psychoanalyst Karen Horney theorized that a woman had PMS because of the denial of her desire to have children. This was yet another example of what Stolberg, the historian, had mentioned earlier: It might not be PMS, but how we choose to interpret it, that is more important.

"In those early years, I had a lot of difficulty expressing myself, believe it or not," my mom said. "When I got my period, I would get angry at all those things that normally I wouldn't. I would get angry."

I didn't say anything, but I would be too if my husband tried to drug me in an attempt to "improve" my behavior.

"I refused to take the Midol sometimes," my mom said, "but when I took it, I realized it was better."

"Why?"

"We got along better," she said. "I calmed down."

I lobbed the next, somewhat facetious, question at my dad. "So when Mom got her hysterectomy, was that a good time?"

"It was awesome," he said.

My mom laughed. They both did. It was a sensitive subject, but I guess they could look back on all of it with humor. Maybe that's how couples make it through the long haul—the struggles, the bumps, all turn into fodder. But then the conversation quickly degraded.

My dad turned the tables. He asked me if I thought I experienced those same mood changes.

My chest constricted. I felt defensive. "Sometimes maybe," I said, "but not regularly." I was telling the truth; it didn't seem to happen each and every month. I noticed it and named it only when my feelings didn't make sense to me, such as the day I went bonkers on Dave that I described at the beginning of this chapter.

Then all of a sudden, my parents started to gang up on me.

"Mara," said my mom. "I've noticed when you get your period."

"You're more sensitive," said my dad.

"Yeah, you are," said my mom.

"How do you know, though?" I asked. "You don't know when I have my period."

"I don't know if Ken knows, but I know," she said.

"How do you know?" I asked.

"Because your manner changes and you have less patience," she said.

This exchange reminded me of being back in high school. They used to sit me down at the kitchen table and recount my mom's PMS story suggestively—the way she'd become incensed and intensely critical. They'd wanted me to reflexively see that I was acting out, just like my mom had done.

See, it's normal. Women get crazy. If you take a Midol, you can calm down just like I did when I had ovaries.

The defensiveness I felt at the moment was the same that I'd felt as a teenager. "So when you told me to take Midol when I was in my teens, was it because you couldn't bear to be around me in that state, or did you want to save me from myself?" I asked.

"I think it was difficult to be around you," said my dad. "I know that's the wrong answer, but you'd be incredibly sensitive. And I actually wanted you to be happier, but it was arguments that were not going to end right—no chance of being resolved—and I'd experienced that and knew all about that."

"And you know what the absolute worst thing you could ever say is?" I asked rhetorically. "'You need some Midol.'"

"It makes women really mad," he said, laughing, and then joked, "That's why Deena wouldn't take her Midol sometimes." He asked me if I ever tried to take the medication.

"No, I never took it," I said. "I thought it was idiotic. It didn't make sense to me. I felt fine."

"But you do get moody, Mara," my mom chimed in. It was funny coming from her—ironic maybe—the same person who minutes before had described how awful it felt to be accused of hormonally incited mood swings.

After more back-and-forth, we slowly shifted subjects—we talked about their dog, Marble, and ended the evening by discussing how perfect this year's crop of avocados was going to be. It wasn't until later, when the neon lights of shops blinked on for the night and I spotted the flickering televisions in apartment windows across the way, that I realized the importance of our conversation.

As a teenager, I'd fiercely opposed PMS—I'd get livid at the mere mention that any of my emotions could be due to some blip in brain chemistry—only to now find myself grown up and embracing PMS as if it were my rightful inheritance. I didn't need anyone to accuse me of PMS now, because I mostly put it upon myself. I used it as an apology and as an excuse. That shift occurred unperceptively and I wasn't sure how or why.

After talking to my parents, I saw just how deeply rooted the idea of PMS is in our psyches. Before I'd even felt a mood shift for myself as a teenager, my parents had already told me that my body would betray my mind. It just seemed like a given that once a month women would lose their shit.

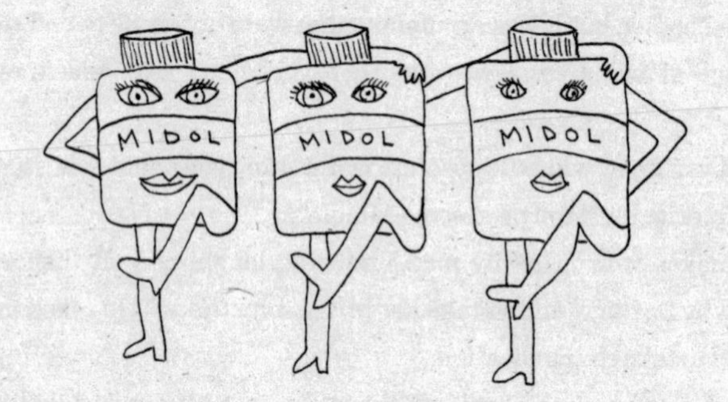

Because PMS seemed so real by this point, I wanted to understand the other side of the argument. Who were the people who don't believe PMS is actually a thing?

I began by reaching out to Carol Tavris. She's a psychologist and feminist, who is well known for her war against psychobabble. She even wrote a book called *Psychobabble and Biobunk* and another one called *The Mismeasure of Woman*. She didn't seem to want to talk, but she sent me an intriguing email.

She listed some studies that proved PMS didn't exist and then wrote, "The 'crying study' is especially cute—women think they cry more when they have 'PMS,' but they don't." Then, as a sign-off, she added, "How much easier to say, 'It's my hormones,' rather than, 'It's George.'"

After that, without even telling me who George was, she went radio silent.

Tavris, apparently, was not into elucidation, so I thought I'd learn more from Joan Chrisler, a professor of psychology at Connecticut College. When I read her many PMS papers (but like, many), I came to the conclusion that she believed PMS was a big wad of horseshit.

Premenstrually, she believes that we have been conditioned to pay attention only to our negative emotions. Even researchers, in her opinion, are biased. Something called the "Menstrual Distress Questionnaire" has been used to assess a woman's mental state. Chrisler wonders how women would perceive themselves differently if it had been called the "Premenstrual Joy Questionnaire."

I got Chrisler on the phone. "So does PMS exist or not?" I asked.

"To ask is it real or not is too simplistic a question," Chrisler said. Maybe it's the submissive in me, but I liked how she gave me the smackdown. "The question is," she corrected, "is it normal or is it a disorder?" She explained that when we say, "I have PMS," we are actually saying that we have a medical condition—an illness.

PMDD (a.k.a. PMS on steroids) is in the *Diagnostic and Statistical Manual of Mental Disorders* (*DSM*). It was inducted under much scrutiny and debate. Detractors, like Chrisler, think it was added in an attempt for pharmaceutical companies and psychologists to reap billions from the pathologization of women's bodies. For extreme cases, antidepressants have become a common treatment.

"If it's really strong and you can't cope with it and you need some kind of help, then you have a condition, but most of us don't have a condition," she explained. "We can just move on." She would prefer if women, instead of blaming their bodies when they felt moody, could just say something benign like "I got up on the wrong side of the bed this morning."

"But me and my friends use PMS all the time," I said. I felt compelled to give her some examples. "We say, 'Today sucked, I must be PMS'ing,' or 'I want to kill everyone—it's probably PMS.'" It wasn't something we shared over megaphones or anything—just within our closest circle of friends.

Chrisler didn't respond, so naturally I had to fill the silence.

"It feels so good," I added, "so how can it be bad?"

"I worry about this a lot," she finally said.

"Why?"

"When we say things like that," she said, "we are saying that our moods are illegitimate."

Chrisler believes that women use PMS as a scapegoat because of pressures to conform to the feminine gender role. “If you want to be seen as nurturing, sweet, affectionate, and a good wife, mother, and friend,” she said, “it’s convenient to have something to blame your stress and sadness and anger on.”

I had to take a moment to let that soak in; it was a trippy and somewhat concerning revelation: By using PMS to rationalize my more extreme emotions, was I actually playing into what society wanted me to be—a demure girl who doesn’t cause a kerfuffle?

All unsavory emotions could be sealed up in the PMS Box and then tossed out—edited out of the story line—like unwanted garbage.

She told me to implant what she was about to say deeply into my consciousness because it was important. “Hormones do not create emotions,” she said, “but they can exacerbate them.”

This was terrible news; I think she was saying that when I think I’m experiencing PMS, I might actually feel the way it feels like I feel.

I wasn’t ready to buckle from her pressure. “But—” I began.

And that’s all I got out before she let out a little sigh. I obviously wasn’t “getting it” fast enough.

Chrisler said that using PMS wasn’t negative only on an individual level, but also for women in general. She made it seem like anytime a woman uses PMS as an excuse for her behavior, a misogynist gets his wings. “A lot of hostile sexists have this notion that women can’t be in any positions of power, because they can’t control themselves and they are just going to ruin everything,” she said. “So they still pull it out! They still use PMS against us! And when we buy into it, we encourage them to do it even though we don’t want them to.”

She had an interesting point, and I felt like a shitty feminist.

“You can’t have it both ways,” she said.

We talked for another bit about the power of labels. I mulled the idea in my head and tried to turn the tables. I wondered what would happen if we started telling men that after they ejaculated, they would suffer from post-ejaculation depression (PED)—emptiness, dissatisfaction, and the desire to flee. Maybe, like PMS, their experiences would suddenly be considered an affliction.

"I love that!" Chrisler said.

She continued to say many interesting things. She's a real majestic, knowledgeable lass, though she'd probably object to the words "majestic" and "lass" because she's so no-nonsense about everything. She's also postmenopausal. She told me it ain't half bad from the other side, so that was nice.

Then as we were wrapping up, she said she was sitting in her office, surrounded by PMS paraphernalia she'd collected over the years. "I'm looking at a sign that's hanging off my file cabinet and it says, 'My greatest fear is that there is no PMS and this is actually my personality.'"

"Funny," I said. And I did think it was funny, even though it also really resonated. Those parts of me—getting misty-eyed at car commercials and being pissed off at something seemingly arbitrary—didn't fit in with the perception I have of myself. "But sometimes," I said, defending myself, "I really don't feel like me."

She reused what seemed like one of her favorite phrases: "Then isn't it convenient to have something to blame that on?"

I wondered what Dave was going to think of Chrisler's theory—that when I shit into his soul, I might actually mean everything I say.

I dug deeper and my mind kept getting blown as I discovered that there were as many theories about PMS as I have colors of underwear (and I have all the colors).

There are too many theories to relay, but a couple stood out. Susan Brown, an evolutionary psychologist at the University of Hawai'i, for example, came to believe that it is ovulation when our mood shifts—we are actually at peak happiness during that time—and that the time pre-period sucks only because we are coming down off a high.

But it was Alexandra Pope's perspective that ended up really speaking to me. Pope is a psychotherapist who also runs an online group called the Red School, which offers courses about the menstrual cycle. She started researching the topic thirty years ago, when she experienced debilitating pain due to endometriosis. She didn't want to take pain medication each month, so she began investigating the subtle psychological and energetic shifts during the cycle and then learned to live in harmony with those changes. Unlike Chrisler, who views PMS as a crutch, Pope sees it as a superpower.

"I reframe the cycle," she told me. Pope, who is sixty-three years old, did not see mood shifts at all in the negative. "If you were the same all the time," she said, "you might safely assume you're dead, really." PMS, to her, is actually an ephemeral moment of clarity. "Your polite socialized self bites the dust, because your energy levels are dropping and your tolerance levels are dropping."

I'd heard a similar interpretation earlier from Ingrid Johnston-Robledo, a board member for the Society of Menstrual Cycle Research. "I think of it as a veil lifting," she said of PMS. "Your need to self-silence is ruptured at that time and you can be more authentic about your anger or your frustration."

I dug this outlook—it was much more liberating than thinking I was out of my mind—but it also brought up some concerns. "I don't feel like it's fair sometimes," I told Pope. When I said that, I was thinking of the moments that I've been extra tough on people, especially Dave. "In a couple of days," I explained. "I know I won't feel the same way anymore."

If all my feelings were going to blow over, what was the point of making it into a big deal?

"That's the criminal thing," Pope said. "You go back into la-la land again." She said we don't have outbursts premenstrually just for kicks. The feelings are real and we need to deal with them—not just pretend it's an emotional blip. "If you come around to the premenstrual time again and you have the same reaction," she said, "it means you haven't dealt with the issue. You are going to sleep again." She sees PMS as a built-in warning system. "We call it the feedback moment, and you don't want to suppress that." If you feel depressed or overwhelmed, that's not just because of hormones—it is because you've actually been spreading yourself too thin. "If you have been putting yourself and your own needs second," Pope said, "this is when you are going to hear about it." She really wanted to drive home that point and spent quite a while berating me with it. "If you're overtired, if you've been pushing it, if you've been eating a poor diet, you're going to hear about it right before the period," she said. "Isn't that brilliant? Isn't that wonderful? Your cycle is actually helping you to take care of yourself and not abandon yourself."

In other words, Pope is saying that if a woman experiences severe PMS symptoms, she's probably been shit at taking care of herself. She needs to lower her stress, sleep more, eat better food, and get a goddamn massage already.

By the end of our conversation, I was enchanted by Pope. I was delighted that she didn't want to deny PMS nor use it to dismiss our feelings; instead, she wanted to highlight it in a positive way and urge women to embrace the shift. In Pope's world, saying "I have PMS" would not mean that you felt crazy or out-of-control; it would actually be a warning to those around you that you meant business.

"You have a more critical eye and the ability to speak the truth," she said. "That's not a vulnerability. That's a power."

She gave me one warning: "It's important to know yourself and your cycle or else during that time you can become quite abusive." She said that it's sometimes best to take note of the issues that arise during that time and then deal with them on a less sensitive day. "Ultimately, I'm saying the truth is a good thing—it's a good thing and we women need to get on board with that!"

I felt like this information could turn the whole world upside down—in a good way.

Clearly, I had to share this information with someone. I called my mom. I didn't relay the theory quite as eloquently as Pope herself did, but I got the point across: "We are not out of our minds when we are PMS'ing, we are speaking our truth."

The change in my mom—the rethinking—happened so fast. Suddenly, she was looking back and rewriting her past. Everything that had happened during fights with my dad—the issues that arose while she'd had PMS—she saw in a different light.

"I did feel really strongly about those things," she said, beginning to validate herself. "I felt like he bossed me around; I was trying to tell him that."

Because she and my dad had been so focused on her behavior, the content of what she'd said was often lost. The same thing was happening to me, but I'd put the blame on myself. I felt guilty about getting upset, so instead of dealing with the issues that emerged, I was left apologizing for being overreactive. "Must be my period, babe," was my tired refrain.

I thought about my anger at the toilet seat being up. I wasn't just being overly critical; I'm a night pee-er. Having the seat up is dangerous. Four separate times, in the middle of the night, my bottom has splashed into the toilet bowl. My annoyance was legitimate.

"I love that so much," my mom said of this new twist on PMS. "It can be like a headlight, showing us how we feel."

By this point, I'd gathered massive amounts of menstrual information. Some of the perspectives I learned about spoke to me more than others, but I still wasn't sure of the reality—was PMS a superpower, a truth serum, a scapegoat, a disorder, or a figment of our imagination?

To find the answer, I decided to immerse myself within a group of hormonal women who were highly attuned to their cycles. By watching, experiencing, and listening to these ladies, I would be able to come to a conclusion. After searching online and reaching out to my period-savvy contacts for such a place, I found Sacred Groves, a retreat for women in the middle of Bainbridge Island, Washington, and signed up for their quarterly Red Tent event.

Red Tents are held in locations all over the world. The purpose is to bring women together to celebrate their cycles. This little island getaway couldn't have sounded more perfect. There would be tons of women under one roof, all focused on shedding their uterine lining. Or so I thought.

While at Sacred Groves, I also planned to try a holistic pelvic session, which is less mundane than it sounds. "It involves putting a finger into the vagina for about forty minutes," Thérèse Charvet, the owner and pelvic practitioner of Sacred Groves, told me over the phone.

I had never—that I could recall, at least—had a singular finger inside my vagina for that long. I grossed myself out by accidentally envisioning a waterlogged finger.

Thérèse explained that by massaging the pelvis (from the inside), she could help release trauma. The trauma could be from something as catastrophic as incest or rape.

"I haven't experienced either," I said.

"We all have trauma," she assured me. "We are living in a male world where the female body is seen as inferior."

The timbre of her voice wavered as if she were speaking into a whirling fan—it made her sound wizard-y. She explained that, as women, whether we are aware of it or not, we've been shamed throughout our lives. "We've been told it's icky, gross, and dirty," she said of our periods. Just by living in a patriarchy, she said, women accumulate trauma inside the body. Because she knew about the nature of my visit—that I was looking into PMS—she assured me that the subject would also be addressed. "We have to learn to honor the rhythms," she said.

Sounded good enough.

I invited my friend Maggie to come along. She'd be a great wing-woman; she's totally go-with-the-flow. A couple of years ago, she even kept her cool when we totaled our rental car in a deserted Mexican swamp. When I told her what I was doing—that Thérèse was going to exorcise the patriarchy from my vagina—she didn't seem surprised.

"Oh," she said, "so she's a pussy reverend."

"Yeah," I said, "I guess."

Two weeks later, we took a plane to a ferry and then drove to the retreat, which was down a dirt road, and then even farther down another dirt road. Cedars and firs, thick with green needles, towered over three circular wooden yurts—two small and one large. They had names like Sun and Moon. Other structures, some haphazardly built with wood and nails, were scattered around the ten-acre property. As we entered, chickens clucked and ran around a small garden. Little altars made up of sticks, shells, statues, moss-covered pebbles, and broken pieces of ceramic were placed at sporadic intervals.

The retreat was run not only by Thérèse but also by her wife, Tere. They were both in their sixties and had a penchant for wearing crushed-velvet shawls in the evening. At first, I was concerned about the New Age vibe. I have a deep and abiding resistance toward the woo-woo, and even though over the next two days they smudged sage constantly, spoke of the Earth Mother, and once blessed me by waving a taxidermy bird wing over my body, I found them both to be smart, surprisingly grounded, and appealingly gritty. I'm not sure I'm ready to admit it yet, but by the end, I may have even come to enjoy beating the "grandmother drum" and talking about my "root chakra."

When we arrived, Maggie put our stuff in the Moon Yurt and I went straight to my pelvic session, which was held across the yard inside the Castle. The Castle, despite its name, was actually an unassuming, tiny two-story structure that was built by Tere.

Later, Tere would tell me that when she used to work in construction, she loved leaving her used tampon in the Porta Potti at the building site. "All the men could see that women worked there, too," she said, laughing, "and even when they're bleeding!"

Maybe I had more patriarchy in my vagina than I'd suspected, because I thought that was clever, yet also gross.

As I walked toward the Castle, I had no idea what was waiting for me. So far in my life, when someone touched my vagina, it was either for sexual pleasure or to open wide for a speculum and a smear. This, from what I could gather, was going to be neither: more like a backrub, but for the inside of my genitalia.

Inside there was a desk, two beds, some astrological-inspired artwork, and a tapestry tacked to the wall. Thérèse sat at the desk near the window; her short gray hair was unconstrained by gels or barrettes. Her outfit was just as casual: jeans and a T-shirt. She had the look of a good neighbor—not the kind who would bring over a casserole, but the kind who'd bring over a handmade herbal poultice from her garden to help draw out puss from a pimple.

She lit a bundle of sage and dabbed the smoke around my body as she told me what to expect. "Some women have memories, emotions, and even colors come up," she said. "Some will get angry, scream, or cry." She also said there are women who don't feel much at all, a vast swath of their pelvic area turns out to be numb inside. "There is something so painful," she explained, "that they just shut it down."

She handed me a 3-D model of the pelvis—without all the organs inside, it looked like a bowl—and told me that's where she'd be poking around. She would massage my muscles not only to relax them, but with the hope of unlocking the underlying emotion that she suspected lay within.

By this point, bony pelvis in hand, I was pretty sure that I no longer had to worry about getting turned on.

"So what's the goal, exactly?" I asked.

She showed me her finger, which would be her tool. "I'm going to sweep away any remnant of shame, fear, and disconnect," she said. Then she stood up and led me to the smaller of the two beds. "In our society, we live from here up," she said, holding her hand to her neck, "we want you to drop in there"—she now was pointing to my crotch—"and let *her* give you guidance and inspiration."

When I got the cue, I took off my pants and lay down. Thérèse sat beside me in a chair, pulled on a blue latex glove, and placed her hand in a loose fist between my legs. Despite my being on an island in the middle of a forest down a dirt road and on top of a flower-patterned bedspread inside a structure named the Castle, the whole thing felt surprisingly legitimate.

"Visualize dropping a root from that part of your body, through the mattress, through the floor, into Mother Earth, and into the ground," she said. "Any releasing we do today will travel the path from your vagina through the root, and Mother Earth will compost it for her growth and for your growth."

I played along. I envisioned a thick root shooting outward from my privates, one strong enough to hold Tarzan.

"I'm at your gate," she said. "Do you feel safe?"

I nodded and then she slipped her finger inside. Her finger was in there for the full forty minutes. During that time, she kneaded my fascia, tested my pelvic strength, and explored the more abstract sides of my vaginal health. Because she believes that we carry the scars and traumas from our family—that they are passed down and stay rooted inside us—she started by calling on my ancestors to help in the process. She prodded my left anterior wall as she said, "Your grandmothers, great-grandmothers, great-great-grandmothers, come help in this multigenerational healing."

I usually try not to think about family while someone probes my genitalia, but I gave it a go.

Then Thérèse began touching different parts—I could feel her up under my sacrum and rapping against the underside of my hipbone. These were places so far up I thought only a surgeon could reach them. I was also surprised, considering she was inside my sex place, about how uncomfortable it was—some spots she pressed made me feel nauseated. There was something deeply perplexing about the fact that I was spending $125 for a vaginal massage that wasn't even 1 percent orgasmic.

"You'd think this would feel better?" I said.

"It's a shame that women think if someone touches their vagina it should be for sex or nothing at all," she said.

I suddenly felt so closed-minded. I was that tourist in Mexico who was upset and surprised that not every meal was made up of tacos and a margarita.

After more time passed, she asked, "What are you feeling now—any emotions or images?"

"Not really," I said, feeling slightly concerned.

"Am I one of those numb people?"

"I don't think you're totally numb," she said.

I didn't scream or shout or writhe, and I feared that my calm and lack of color associations were disappointing Thérèse. I wanted to come up with something we could work with. I closed my eyes and tried to associate. Unfortunately the first thing that came to mind was ridiculous. "I feel like your finger is skiing inside of me," I said.

I mean, it did; she was moving the digit along my fascia like it was a snowy black diamond. "Your finger is downhill skiing," I said again with more conviction.

She didn't say anything, so I got self-conscious.

"Am I doing this right?" I asked.

"Yeah," she said, "tell me anything you're thinking, because it might be clues about what you're holding in here. Ever had a skiing accident where you landed on your bowl?"

I didn't expect the translation to be quite so literal.

"No," I sighed. If these vaginal walls could talk, they would apparently bore people to tears. But then, I suddenly remembered something. I never could have imagined I'd be so excited to reveal this little factoid. "I did land vulva-first on the high beam before!"

When I was a gymnast, it was unfortunately one of my most popular moves.

"Oh good, that might be part of the tension," she said. "Light early injuries can make a mark." She began swishing her finger back and forth around the circumference of my bowl. "Think about banging your crotch," she said, "and send that down your grounding port."

After I sent all my vulva thumps down the "grounding port," we worked on sweeping the shame and fear away—all the stuff that we as women accumulate up there. "All the fear is going down through your root," she said, as she mimed pulling something down and out through my canal, "and into the ground."

We spent the rest of the time trying to facilitate a release of tension because we found out that I'm vaginally retentive—my muscles are tightened almost all the time. "Relax here," she said, wiggling her finger. I felt vague pressure near my rectum. "Breath into it so we can soften it up."

I breathed and released. She moaned and asked me to moan along as if I were in labor and trying to deliver my vaginal knots. "Ah, there we go—there's much more space in this pocket now," she said.

At the end of the session, she said my bowl had become much more malleable. She said the blood flow was stronger and my muscles responded better. "This would be so much healthier for women to do than a Pap smear," she said. She views this type of massage as preventative medicine. She pulled out her finger and said some blessings, then sat by my side as my eyes fluttered open.

I was relaxed and peaceful, but I still couldn't help getting back on topic, the one I'd originally traveled there for. "So, um," I asked, "how is all this going to help me figure out PMS?"

"It might not," she said.

I gave her my confused look: duck lips and furrowed brows.

"But it's getting you more connected to that part of your body," she said, "and that's important for all of us."

She stood up, grabbed some papers from her desk, and turned out the light. I was still lying down, wearing only my shirt. "I encourage you to spend some time alone with your bowl," she said as she walked out.

The Red Tent wasn't until the next day, so Maggie and I spent the afternoon exploring the property. As we hiked through the forest, she asked how the pelvic session went.

"I feel like I got the wood floors refurbished," I said. "Know what I mean?"

She liked the imagery—she's all about home renovations—but wanted me to further elucidate.

I didn't expect to like it—it seemed a little out-there—but when I'd stepped out of the Castle afterward, the world felt more vibrant. I was high, but without the dry mouth. I saw a brown and beige chicken clucking, and a huge-ass grin ripped across my face. For a fleeting moment, everything made sense.

It was all so simple: *I am hungry. I will eat something and not be hungry anymore.*

I was convinced that every woman should be able to go down to the local strip mall and get her vagina muscles tenderized with the same ease she can get her nails done. The world would be better for it.

Then Maggie farted. "I think I just had an emotional release," she said, making fun of me.

That evening, Thérèse and Tere were hosting a wailing circle. I didn't sign up for it, but it was happening anyway. From all over town, people came together to grieve. They could grieve about whatever they wanted—loved ones who had died, the state of the planet, traffic jams, a hangnail. The idea was to give people an opportunity to emote in a society that is often considered cold and emotionally remote.

The circle took place in the large yurt, which was decorated exactly how you'd expect a home for women who would host a wailing circle and a Red Tent at their residence to be decorated. The sofas were mismatched. Feathers, candles, geodes, and shells made up 75 percent of the interior design. No matter what direction you looked, there was a painting of a woman. There were mugs on every flat surface and enough boxes of tea in the kitchen to build a pyramid large enough to fit their overweight cat, Oreo, inside.

Tere beat on the grandmother drum, as twenty-three people walked in a circle, screaming, crying, and falling to their knees. The whole thing made me uncomfortable. I'm not a griever. I have been properly trained to hold all my pain and suffering inside. I excrete my emotions the polite way: in ulcers.

The good thing was that after the grieving ritual, I was able to ask one of the attendees about her PMS. Her name was Jude. She had long white hair and was now postmenopausal, but she said that in all her years of menstruating, she'd never experienced any symptoms. "I think society builds that tension in people," she said.

She explained that she'd dropped off the grid and lived with her husband in the hills of West Virginia. She spent most of her time frolicking naked in the woods. "I was never stressed out," she said, "so I didn't have to release anything—I wasn't living in a way that made me unhappy."

What she said reminded me of a study I'd read: Despite having hormonal changes similar to Western women, the !Kung, a hunter-gatherer people in Africa, never associated the menstrual cycle with a change in mood. The same went for Samoan women and the Rungus of Borneo. While the physical symptoms—cramps, lower back pain—seemed near universal, Alma Gottlieb, a professor of anthropology, reported that the psychological aspects of PMS were exceedingly Western.

Later, on the way back to our yurt, Maggie threw down some knowledge from her acupuncture studies. She told me that in Chinese medicine, PMS is due to liver qi stagnation. She says weird shit like that all the time—like when I have a stomachache, she'll ask me if the pain feels hot or cold.

As usual, I had to ask her to translate.

"If you have PMS," she said, "we say that you're getting blocked up, because you're not getting something you want—you're not getting your deepest desires met."

This was odd; she was someone who had always blamed her bad days on PMS, on her brain chemistry gone haywire.

In light of that, it was surprising to hear her say it was due to something deeper. "So," I asked, "what do you think your PMS has been trying to tell you?"

She thought for a moment and grew increasingly serious. She stopped and looked around at the towering firs. "I guess, a lot," she said. "I feel insecure financially and am deficient in taking care of myself." She admitted that most of all, she probably needed to concentrate on bringing more love into her life. "I want a partner," she said as pine needles crunched under our feet.

In the morning, the sunlight funneled through a big circular window at the top of the yurt. I slowly became conscious as the sky lightened minute by minute, hue by hue. It was so peaceful; it made me think of Jude. Maybe there was something to her theory—if there was no stress, maybe there would be no PMS.

In other words, if the toilet seat is not constantly up, then you can't get upset about it not being down. But then I turned over and saw a black-winged ant the size of a thumbtack on my sheet. I flicked it off. About two minutes later, it was crawling up my pillow. Then Maggie awoke to one clinging to her bedspread. She took a shoe and smashed it. I flicked away two more and saw one fly into my suitcase before we retreated to the main yurt.

"Did you meet our anties this morning?" Thérèse asked. She sounded close with these insects.

"Were we supposed to take care of them," I asked, concerned, "because I murdered a few."

She told us that they had accidentally built the yurt on top of an ant's nest and they'd done everything possible to try to get rid of them—poison, sucking them up a vacuum, and even praying. "Then I went on a vision quest," she told us, "to try to speak with the ant queen."

"You talked to the queen ant?" I said, confused. It was still early and I forgot that I was in an alternate universe.

She told me she did it through a shamanic journey. "So you made some peace overtures?" said Maggie.

"Yeah, but the queen ant was like, 'You're not the great Earth Mother, I am, and now you're trying to kill me, what a hypocrite you are!'"

Thérèse said the queen ant told her that the only way she'd leave was if they burned the yurt down. "So we backed off and now we try to make peace with them," she said.

Since we were already halfway through the weekend and yet to speak of PMS, I wondered if Thérèse was making a subtle metaphor.

After breakfast, it was finally time for what I'd come there for: the Red Tent. We constructed a literal one in the living room. It was made out of a metal frame with different swatches of red fabric tacked over the top and sides until it closely resembled a street-fair booth that would have Middle Eastern rugs for sale inside. By the time everyone arrived, it was one p.m. and there were fifteen of us spanning from age eleven to seventy-three.

Susan, the eldest, showed up first. "I'm glad we are living in this day and age," she told me. "If we lived in the colonial times, we'd be hung." That seemed like an extreme outcome for our actions thus far; all we'd done was put out some tubs of hummus. I hoped her comment meant something outré was about to happen. Was Thérèse about to command us all to strip down and paint friezes with our menstrual blood?

Before Susan could elaborate, the cat distracted her. "Come here, Oreo!" she said, ambling toward him. "Oreo!"

Maggie and I milled around, visiting the snack table multiple times.

Mary, a woman in her forties, washed some dishes. Her two daughters were on a sofa, chatting with each other. Another woman was sitting with a book by the window. A mom and daughter from Austin, Texas, had gone for a hike somewhere outside. At that point, it was already close to two p.m.

I asked Thérèse when the Red Tent was going to begin. "It's already happening," she said, smiling and opening her arms toward everyone in the living room.

This was what I was anticipating all weekend? People would get hung for *this*?

The way everyone was loitering reminded me of the waiting area at an Amtrak station. But if that was indeed the case, I tried to make some inroads and deployed Maggie to help me drum up some PMS conversations, but no one seemed particularly interested in talking about their cycles, which was odd, because I thought this whole thing was supposedly about how we bleed through our crotches.

I felt concerned; I had traveled all the way to Bainbridge Island, Washington, to become enlightened. "When are we going to talk about PMS?" I asked Tere as she prepped chia-seed pudding in the kitchen. I noticed her mustache—how she didn't mess with it, but rather kept it dark and conspicuous.

These ladies are so cool, I thought, *being themselves in the middle of the forest.*

Tere said we'd address PMS later, during the ritual portion of the evening. "For now, relax," she said, "partake."

At first, it didn't seem like much to partake in, but I soon found myself submerged in a galvanized horse trough that Tere had transformed into an outdoor hot tub. Five of us, including Maggie, sat in there, proudly bearing our breasts and bushes under the sky, under the birds chirping, under the fresh cold air. We managed to get lost in conversations about death, babies, and career. There was also a heated debate about coffee dates. "You shouldn't say yes to coffee unless your intention is to actually go," said Mary.

"But how are you supposed to say no when someone stands before you and says, 'Let's get coffee sometime'?"

"You have to be authentic," she said.

"And hurt someone's feelings?" I asked.

Besides, I thought, "Let's get coffee" had simply become a figure of speech. People said it so often with such little conviction that it had become an alternative punctuation; it was a synonym for "goodbye."

After every square inch of my body was waterlogged, Thérèse walked with me to the Red Tent. She swaddled me in a blanket and placed hot rocks that had been heating in a slow cooker on my chakras. She covered my eyes with a piece of cloth and turned up the music as I took in gentle wafts of her chamomile breath. As I fell in and out of sleep, she whispered blessings into my ear.

"Bless your mouth and throat that you have an open channel to speak the truth," she said. "Bless your feet—may they always feel their connection to Mother Earth's support and ground."

Four other swaddled bodies already lay on the ground around me, and I could hear the steady breathing that signified sleep.

Thérèse tucked the sheet a little tighter around my chest. As she left me to rest, she said, "You're perfect."

By dinnertime, I felt completely immersed in the moment and apathetic to my original quest. People were scattered around, eating in little groups. I sat next to Tere, whom I'd really come to like. She was lanky and had short brown curly hair with an androgynous appeal.

As our plates teetered on our laps, I pointed to her mustache. "So is that like a 'Fuck you' to society," I asked, "like you're not going to play by their rules?" I thought that was one of the modus operandi of lesbians who lived out in the woods.

She looked at me—opened her eyes wider—and then crooked her head to the side. "No," she said. Then she touched her upper lip with her fingers. "Is it really that bad?"

I shook my head. "No," I said, stitching my brows. "It's totally fine."

She moved some salad around her plate.

"I just thought," I said. Then paused. "Never mind."

We went back to picking at our vegetarian enchiladas, but then I stuck out my chin and told her I had some whiskers. Even though I'd come to terms with them, I still did my best to hide them from the world, so what I did next surprised me. "Do you want to rub them?" I asked. She rubbed mine and then she stuck out her chin so I could give hers a rub, too. Tiny grains of stubble tickled my fingertips. When we were done sharing, I said, "I've never done that before."

"Me neither," she said.

We laughed as we cleaned off the last bits of food on our plates.

I had a fleeting feeling that it was actually moments like those that the Red Tent existed for.

Finally, at seven p.m., everyone gathered for the ritual part of the evening. We formed a circle inside the tent—some sat on the maroon sofa, others on big pillows, and a few on the wood-slated floor. For the next hour, we meditated on our ovaries and then shook maracas while we sang songs with lady-centric lyrics.

Oh, Momma Ocean, hold me close and tell me your ways
I feel your tides inside of me
I surrender and I am set free

It was all as corny as a fifty-year-old man with high-waisted Dockers telling knock-knock jokes, yet I still managed, when guided, to earnestly visualize my egg traversing my fallopian tubes. Then, as the evening began to wind down, I finally came back to myself. I realized that there had still been no mention of PMS. I resolved to hijack the group and appeal to the women myself. "What do you all think of PMS?" I asked.

It was quiet for a moment, but then a few women piped up.

There was the girl, Olivia, who said that she acts out. "I just warn my friends," she said. "I say, 'I'm PMS'ing, so I'm going to be a bitch this week.'"

I thought about Chrisler—how she thinks that we use PMS as a permission slip to show the gruffer side of ourselves.

Then another woman said, "I'm so sick of having shame for being pissed off."

But then the conversation quickly petered out. Thérèse was standing at this point, getting ready to wrap everything up for the night.

I looked to her, somewhat accusatorily, and said, "Why didn't we talk about this more?" I was pissed. Before I came, I made sure that PMS would be heartily addressed.

"This is the woman's way," she said with no anger or ill will.

That made absolutely no sense to me, but it turns out that didn't get in my way of accepting it. I was so content from a day spent lounging that, at that point, her answer—maybe the fact that there was an answer at all—seemed perfectly satisfying.

"Oh, right," I said, nodding. "The woman's way."

Everyone got up and Mary said to no one in particular, "When do you get to let go like this?"

We all hummed in agreement and then parted ways as we went off to sleep.

I woke up at six a.m. with the chickens clucking and the sky just beginning its transition from black to blue. I felt calm and relaxed until I stared up at the ceiling. Last night I must have been in a chia-pudding-induced coma. *What did that even mean—"the woman's way"?* Maggie and I would be leaving in a few hours, so I threw off my covers and rushed to the main yurt.

Tere was alone in the kitchen, brewing some coffee and tipping a bag of oats into a large water-filled saucepan. "'The woman's way'?" I said. "Explain that, please."

Tere poured herself a cup of coffee and gave me one, too. She told me that where the masculine way was direct and goal-driven, the feminine way was circular and open. "You put your question out there, which you did before you started the weekend," she said, "and then you sit back and wait."

"You just wait?" I said. "That sounds unproductive."

"It's not generally productive." She laughed.

She said that the masculine style is accepted by society and is what we think of as "productive"—it is to narrow your vision until your goal is accomplished. "But the woman's way," she said, "is, 'I don't really know what is going to bring me the information I want and there is no way I can direct it, but I can open myself to everything that's coming at me and learn and somehow what I receive will be greater than my original question.'"

I tried to parse what she was saying.

"And that state of not knowing—when you don't have a fucking clue—is also when you're closest to God, when you're closest to the creative force," she continued. "And that's a hard place to be for us humans, because we really want to know."

"So PMS?" I said, shrugging.

She stirred some nuts and dried fruit into the saucepan. "Yeah," she said, "embrace the chaos."

A few weeks after the weekend at Sacred Groves, on a Saturday morning, Dave and I went out to breakfast. We were walking to Pig and Khao, a Southeast Asian place, on the Lower East Side. I'd been building up to this conversation for a long time—since back when I learned about (and was permanently traumatized by) puppy fumigation. We were in lockstep, crossing Houston Street.

"You know," I said, "when I'm PMS'ing, I actually mean everything I say."

"I know you do," he said.

"You do?"

We were at a stoplight, waiting to turn down Clinton Street.

"Yeah," he said, "you make that abundantly clear, but you're at a ten instead of a four." After we crossed the street, he said, "I like it better when you're at a four."

"If you listened to me," I said, "then maybe I wouldn't have to become a ten." I told him about the spool and how it unwinds—all my frustrations get the green light and are released.

"Not every relationship has to be perfect," he said.

We reached the restaurant at this point, but stopped just outside.

I grabbed his arm and turned to face him. "But look, I want you to put the toilet seat down," I said. "It's really important to me." I was so calm that he couldn't possibly deflect what I'd said, nor could he tell me that I was shitting into his soul.

"I'm not being malicious," he said. "Honestly, I just forget."

That was the first time we got beyond accusation and evasion.

I took in what he said and understood how forgetfulness could be the culprit: he didn't meticulously plan out and desire his wife to fall—tush first—into the toilet bowl. How could that be a benefit? We share the same sheets after all.

We took two seats at the counter, and while we shared a plate of longonissa sausage and fried eggs, we talked about how we might be able to fix his memory. When we got home, I enacted our plan. I took out a piece of eight-by-eleven-inch paper and a Sharpie. I wrote on the paper in big block letters and then taped it to the wall behind the toilet. The sign said, "PUT THE FUCKING SEAT DOWN."

We continue to live in imperfect harmony, but more often than not, my butt stays dry.

Wart, Me Worry?

I am five feet tall, yet my shoe size is an excessive 8½. I was built with the foundation to support a skyscraper, but only a small mobile home was docked atop. By the time I was twenty, I had bunions—bone nuggets that jutted out at my big-toe joint like toadstools from a crack in concrete. At twenty-two, a doc lopped the suckers off and cranked my big toes straight. Now, on the top of each foot, I have jagged Frankenstein scars. The nail on my pinky toe, which is the unseemly size of a cricket's eye socket, regularly falls off. All it takes is the friction of putting on a sock. Basically, if I were Cinderella, my glass slippers—if they had any chance of fitting—would need to be made of Teflon and look like two wide gravy boats.

My feet have issues: Because of their relative girth and gnarls, I call them peasant feet, which I pretend is prized in this farm-to-table atmosphere. I've handled all of the above with grace, but there is one issue I haven't mentioned so far, one that was harder to deal with than the rest, which was my stint with foot warts. This is the twenty-first century—isn't it time we see a Disney princess who has extra-wide feet spotted with viral growths?

"No," you say?

Okay, you're probably right.

Foot warts (known more widely as plantar warts) are a strain of the human papillomavirus (HPV), just like genital warts, but instead of colonizing your junk, these little virions have a thing for feet. At least if you have genital warts, there is some likelihood that you had good sex to get them. With foot warts, on the other hand (or foot), there is absolutely no chance of orgasm. On the upside, foot warts rarely cause breakups or the need for therapy.

I spoke with Adam Friedman, an associate professor of dermatology at the George Washington School of Medicine & Health Sciences in the hopes that he could convince me that warts aren't as grotesque as we've all been lead to believe and that I could, with extra information, chalk them up to being just one more rustic characteristic of my artisanal feet.

The professor began on a considerably anticlimactic note. "A wart, plain and simple from a visible perspective," he said, "is extra skin."

He explained that for the virus to slip inside our bodies, we must have a micro-abrasion. The abrasion can be so tiny that we wouldn't even notice it, like a patch of dry skin. Once inside, the wart virus weasels its way to the bottom part of the epidermis. It then inserts itself inside our skin cells. "Which is why warts can persist for so long," Friedman said. "They are living within us, so to speak, and our immune system just doesn't pick up on it."

While inside, the wart doesn't just chill on the sofa like a houseguest. The wart is much more intense; it wants to take over our lease.

"They insert themselves into the various machines inside our cells and start making the skin turn over faster," Friedman said. A wart looks like a really thick callus, he explained, because it's skin that's been making itself way too quickly. "The difference here is that this new skin is chock-full of wart virus with the ultimate hope to spread and multiply."

Getting a wart, in other words, is our body's microscopic remake of the film *Alien*.

People have been trying to cure warts for much of human history. As far as I can tell, no one has ever been like, "Oh, look at that barnacle on my

extremity, let me nurture it." No, they want to get rid of them so badly that they've resorted to all manner of crazy.

In a paper called "'Warts and All'—The History and Folklore of Warts" published in the *Journal of the Royal Society of Medicine,* author D. A. Burns catalogued many of the outré methods proposed anywhere from Roman-Hellenistic times to the 1800s. In one case, wart sufferers were told to rub a slug over their warts for nine consecutive days and then impale each slug on a thorn. "For as the snail, exposed to such cruel treatment, will gradually wither away, so it is believed that the wart, being impregnated with its matter, will slowly do the same." Another "cure" involves a dead cat, and yet another uses fasting spittle. "Spittle must be early morning, pure and unadulterated," wrote Burns.

Morning spittle is clearly the only way that spittle makes sense.

We can laugh all we want at the slug murderers, but to this day, we still have no foolproof cure for warts. Friedman said that plantar warts will often resolve on their own after a year or two, but even when doctors go the extra mile to freeze, burn, or even surgically remove them, they may not be gone for good. "The thing people don't get," he said, "is that the virus is likely living inside the normal skin cells around the wart, but it's kind of asleep." He explained that the leftover virus, for a number of reasons, could reawaken and cause a resurrection of the wart. He also said we should never scratch our warts, because that, in a lot of cases, is what causes them to spread.

When he said that, I admitted to him that I suspected I had a scalp wart, which I often scratch at when I'm bored. "I do all of the things you're not supposed to do," I told Friedman, "yet my face hasn't gotten warty yet."

"I don't recommend doing that," he said. "That's a terrible idea. You can play wart roulette if you want, but you're right, it's also not a hundred percent that you'll get more warts."

His response concerned me. I never do well at the slots.

Dermatologists still aren't sure why some people are more prone to warts than others. "It probably has something to do with our immune system and genetic predisposition," he said.

With that in mind, foot warts, depending on the person and circumstance, can and do spread widely. All it takes is for one infected skin cell left on the ground to be stomped on by another bare-footed person who has the most imperceptible scratch on his sole. One of my friends—let's call her Deb—told me that she gave foot warts to her boyfriend. I asked and received permission to call the guy up to see how he dealt with the transmission.

"How did you feel when Deb told you that she had foot warts?"

"I didn't care," he said.

"And you knew they were contagious?"

"She kept saying that, but I was like, 'Nah, I'm not going to get them.'"

"But then you got them."

"I did." He said he had about four now, which is more than Deb ever had.

"Only one of them is like, 'Fuck you, I'm staying forever,'" he said, "but the others aren't very serious."

I asked him if he was upset about his new additions, but he was really chill about the whole thing. "Deb was mortified, but my reaction was, 'Aw, I have your warts,' which displeased her, but I treated them like they were *ours* and they were adorable."

I found this wart story endearing, but I wasn't impressed. Ultimately, I'll have to stand by my initial sentiment. Warts are gross.

Epilogue

I was in the driver's seat of a black Honda CRV, heading north on the I-15 with my mom in the passenger seat. We were on our way to Glen Eden Sun Club, a resort community located in Corona, California. This wasn't any old resort, though; this was a nudist resort. Not clothes optional. Naked only.

At first, the resort wouldn't let me come for a visit. If you're married, they want both spouses to come. When only one has come in the past, it has bitten them in the ass. "Husbands have shown up at the gate agitated and jealous," Jane, the receptionist, explained.

But Dave, I told them, would undergo waterboarding before he'd go to a nudist resort. He'd take a bullet in the knee before waving his schlong around the 155-acre rural swatch of land, which was a problem because I needed to be around naked people. After spending the past year plus investigating individual body parts, I wanted to experience the body as a whole.

After obtaining a notarized letter (that the resort asked for but never ended up taking) from Dave explaining his position—"I'll eat raw road kill before you get to see my hairy balls"—and five more phone calls, Glen Eden finally granted my mom and me permission to come for the day.

There was another issue, though. My mom wasn't the most gung-ho participant, either. Earlier that morning, I'd come to pick her up at her home, but she wasn't ready yet. She was dithering while packing up a bunch of clothes. "But we're going to a nudist resort," I kept telling her, as she stuffed sweaters, workout gear, wool socks, and a big fluffy robe into a large blue tote. Her overzealous preparation was either out of habit or hope; I was not sure.

She was also running late because she'd done her hair. When I asked her why she put forth the effort, she explained that usually she can dress nice to make a good impression, but that's not an option when you have to be naked. "It's the only thing on my body I have any control over," she said, fluffing her 'do.

She had done the nudist thing as a twenty-one-year-old and remembered it being a sweat-provoking experience. Forty-four years had passed and she lamented how time had treated her body. "Do I really need to do this again?" she seemingly asked a woven shawl, before throwing it into the tote as well.

Because of my persistent harassment—"Mom, whether you like it or not, this book needs to end with two generations of women embracing their female bodies, so let's embrace, okay?"—she found herself next to me in the car at ten-thirty on a Saturday morning in early April, bound for a nudist resort while armed with more clothes than one would pack for an Alaskan sojourn.

After an hour drive, we pulled into the Glen Eden parking lot, a small patch of asphalt beside a brown trailer that serves as office and reception. The resort is just off the freeway, but hidden down a narrow tree-lined road and nestled between a ring of scrub-filled hills. Suburban tract housing and strip malls were only minutes outside the premises, but seemed to exist on an entirely different planetary plane.

As we walked toward the office, my mom pulled her sweater tight. "It's so cold," she said. I rolled my eyes like any compassionate daughter would and said, "Come on, it's not that bad!" I'll admit that it was cold, but it was California cold. It wasn't New York cold. It was in the sixties with some scattered clouds and a slight breeze. I tried not to worry. Nudists, I assured myself, surely wouldn't be practical people. If they were true to their agenda, they'd go nude no matter the climactic obstacles.

The two women at reception were clothed, and I got concerned that because of the weather, everyone would be dressed. But then I learned that clothing was the rule for front-office staff. They gave us papers full of regulations. Two that stuck with me: No PDA (this was no swing club!), and carry a towel around with you at all times to place on any surface before sitting. *How totally appropriate and respectable,* I thought, of how they broached the possible transmission of ass germs. I happily signed my name on the bottom line.

The final step before being granted admission: The ladies hovered over us as we placed an orange circular sticker over our phone's camera. "We had a man who recorded people for a year," one said, "so no more of that." She told us to wait outside for our tour guide, who'd give us the lay of the land.

A few minutes later, careening toward us in a golf cart, came Sheryl, a woman with a boyish haircut and an athletic build who'd lived at Glen Eden for more than twenty years. "You gals are lucky," she said. "I usually make everyone take the tour naked, but it's chilly today."

I looked up at the gray sky, the leaves blowing across the path, and the goose pimples on my arms. "It's not that bad," I said, somewhat desperately. "Let's do it naked anyway."

"Too cold for me," Sheryl said, zipping up her windbreaker a tiny bit more.

My mom raised one eyebrow and gave me a smug smile. Nudists, it turns out, were infuriatingly sensible.

We climbed onto the golf cart as Sheryl pumped the gas. The narrow looping lanes were packed with mobile homes and RVs, each only feet from the next, but bedecked and made unique with flowering plants, tchotchkes, and endearing signs like "Life Is Short, Party Naked" and "Caution: Menopause at Work." At least two hundred people live in the community permanently, while more than sixteen hundred members visit for day trips. It felt like an adult summer camp with its many facilities—tennis courts, pickleball courts, an art and sewing studio with the sign "Naked Needle Gang" hanging in front, a café, a rec hall, a gym, an indoor pool for water volleyball, and the outdoor pool area, which was our final stop. "Usually, you have to come at six a.m. to get a lounge chair," Sheryl said. Then she looked up at the sky. "Not today."

There were more than two hundred chairs. Only two were taken.

She said that because of the weather, she'd allow us to be partially clothed that day. I was disappointed in her willingness to make her guests comfortable. "Have fun," she said, before disappearing inside the Sunshine Café for lunch.

My mom, trying to contain her excitement, changed into her massive white robe and lay down on a lounge chair. With no secure closure and the connotation of pre- or post-shower wear, the robe gave the illusion that she was more scantily clad, but in actuality, she was bundled in about four times more material than she would be on a typical day. I could see only her big toes and eyeballs. I, meanwhile, remained resolute. I would experience a nudist resort.

I shimmied out of my clothes. I didn't have the same trouble disrobing as I did months before when I'd ridden through New York City topless, even though it would make sense if I did. See, since the bike ride, massive bodily changes had occurred.

I was now seven and a half months pregnant with twins.

Many of the bodily issues I had before—the ones that had caused me to talk to scientists, sociologists, anthropologists, and strangers from the casual encounters page of Craigslist—had been exacerbated by the two fetuses homesteading in my abdomen. I used to think I had hemorrhoids, but those ones were for amateurs. I now could swear I had pool floaties where I once had a butthole. My belly button went from a mature outie to a gigantic protruding stump. The hare could have a comfy nap on it while waiting for the tortoise to catch up. And not only had my goatee filled in even more, but I also grew a few black chest hairs conspicuous enough for Dave to think it perfectly appropriate to pluck them, using pointer finger and thumb, when we recently found ourselves in direct sunlight. His actions made me tear up with an intense mixture of emotion—shock, shame, but also an unexpected dash of delight. Though horrid, it was also intimate and accepting. In a giant upset to my preconceived notions, if I was ever in a coma, it might actually be my husband who maintains my beard.

I spread a towel out on the lounge chair next to my mom and plopped down. The chill in the air was refreshing, I swear. "It's a whale," said my mom.

"What is?" For a moment, I thought she was making fun of me in my new ample form.

"That cloud," she said, pointing.

She was not on task at all. She lifted her pointer finger again. "And that's an alligator, for sure."

While fighting to access this place for the past two weeks, this was not even close to the most far-fetched scenario I had fathomed—naked and alone by a pool while my mom played Rorschach with cumuli.

But then those clouds opened up, albeit a very small amount, but enough for people to begin emerging from their trailers and wherever else nudists hide in the cold. I was happier to see other people than I was to stop shivering. There was a dude who looked like a badass Hells Angel; he wore a white T-shirt under a leather vest and big silver cuffs in his ears, and had a salt-and-pepper beard down to his chest, but also, no pants. I hate to say it, but seeing his little tush made him about 100 percent less intimidating. He met up with a group of guys under an awning to play a board game.

Meanwhile, a half-clad man hit volleys on the tennis court while a woman, with only a visor on, drove her golf cart toward pickleball.

Then, to the left, we heard a sudden bout of live music. We walked toward the sound and saw a group of about fifteen people, all in various states of undress, serenading one of the mobile homes. "That's our ukulele jam group," explained a woman to my right. She was wearing only a sarong around her waist and banging a tambourine along with the beat.

They were singing "You Are My Sunshine" to a resident who had just turned ninety years old. A sign hung from her balcony: "Come by for cake from 11 to 1 pm."

Before the hour was up, there were at least twenty people enjoying the atmosphere around the pool without a scrap of cloth for comfort, protection, or concealment. Two guys bro'ed out as if at a neighborhood barbecue, except they were in the Jacuzzi and completely nude. "I'm Ernie," one said, reaching out his hand. "Herman," said the other, shaking hands. They cracked open some Millers. Before long, they were talking about Nam. Both, it turns out, had been drafted around the same time.

"I like the types I'm seeing," my mom said. "They let their cottage cheese show."

My mom has never been one for political correctness. I'm not usually either, but in her presence, I seem to get a little bit more precious. "Yes, there's a lot of different body types," I both agreed and corrected.

You can't stare that long, because someone will call security (this rule was clearly stated on the forms we signed during registration), but I took peeks here and there. The majority of the people were over fifty and had figures that are typically shamed, ridiculed, or deemed undesirable. There was dimpled skin everywhere. Scars from surgeries past were paraded around like a stubborn spaghetti stain on a beloved old couch. There were thick thighs, tiny cocks, bunched-up tummies, hairy moles, wrinkled bums, rolls of flesh, and breasts long and droopy, like overstretched taffy. And no one so much as flinched or tried to hide.

This was a place where people came not to perseverate on their imperfections, but to revel in the fact that they had a working body—the most miraculous thing of all.

As we went to order some lunch—a hamburger and a BLT—my mom finally took off her robe. Was it the weather or the diverse company baring themselves that had freed her from the ten pounds of terrycloth? I couldn't be sure.

We ate by the pool and then went for a swim. Nudity, after a few hours, had normalized; it became background noise to a perfect relaxing day. We'd stopped gawking and cocking our heads at new arrivals and instead gossiped about people we knew. My mom was also sure to sprinkle in some platitudes about what to expect of motherhood. "Sure it's difficult, but it's also the most amazing, life-expanding thing a person can do." I'll remain cautiously skeptical until I meet these little humans. Huge life changes, I've found, always turn out best when first met with low expectations.

Many of the people around the resort also commented on my belly. "It's so great to see a pregnant woman enjoying her body," said a man in his sixties with a paunch to rival mine. To nudists, it seems, the definition of "enjoying one's body" meant baring it all in front of strangers. I enjoyed the term's specificity in this context and wondered if they thought it at all possible to savor their own form with khakis on.

Before we left for the day, we tried our hand at pickleball. By that time, six p.m., everyone had deserted his or her athletic pursuits for the potluck and live DJ happening down at the rec hall, so we had the place to ourselves. We slipped on our tennis shoes and went to the courts. With thirty-five extra pounds, I wasn't tremendously agile, but I did my best, which mostly meant standing in one place while my mom ran around collecting balls that I couldn't run and hit.

Before we left, I needed to do one more thing. I surreptitiously peeled off the orange sticker and handed my phone to my mom. I went deep and held up my racket as if about to hit a ball. I wanted a photo to remind myself of everything—the hair, the sweat, the swollen ankles, the acne covering both face and chest, and the massive ball of babies on my front—because in that moment, that's who I was and there was no shame.

Acknowledgments

The many academics, researchers, and specialists who took time out of their busy lives to share their expertise with me—thank you from the bottom of my plantar warts to the top of my head lice. Without their generosity and knowledge, this book would have been very uninformative! I'd like to thank my bff Chandra Lee Breslow, Reyna Texler, Nora Weinberg, Brian Abrams, Natasha Chey, Luke Kummer, Nita Praditpan, Gabrielle Galanek, Josh Breslow, Margo Adler, Jen Weiss, Jana Winter, Karen Kashkin, Judy Altman, Phoebe Schraer, Esther Altman, Julie Hassman, Briony Chown, Mariah Dilworth, and Maeve Higgins for their support, encouragement, and/or body-related musings.

Warm breezes, Williamsburg Bridge walks, lox-and-bagel sandwiches, expansive views (looking at you, Brooklyn Bridge Park), Chucho, and cottage cheese (the dairy as well as the thigh product) for sustenance and/or inspiration. They aren't traditional writers' colonies, but I'd like to thank the coffee shops—Holsem, Caffè Calabria, and Lazy Llama—that offered me space and delicious brew while I contemplated, freaked out, procrastinated, and finally put words down on paper.

The most gargantuan thanks goes to Kerri Kolen, my editor (as soon as I heard her last name, I knew we'd make the perfect team). She's been a dream come true—attentive, patient, enthusiastic, and opinionated about hot-button issues like vaginal discharge and earwax. Thank you for believing, for giving essential feedback, and for being gross. Thank you to the team at Putnam, and to Anabel Pasarow, Margo Lipschultz, Elena Hershey, and Brennin Cummings for picking up the reins and treating this humble hemorrhoidal book like a Triple Crown thoroughbred. I'm indebted to copyediting mavens Maureen Klier and Anna Jardine for their meticulous reading and their grammar wizardry. Adil Dara Kim and Leah Goren for creating an eye-catching jacket that shows we can be glamorous and hairy at the same time. Erin Hosier, my agent, who believed that pubes deserve a place in American literature and who went out on a limb to help me help them claim their rightful position. Also, for still answering my calls. I don't take that for granted! Maggie Weinrich for being the absolute biggest support as well as a gung-ho sidekick during the majority of these bodily adventures. You're the best, Mags! Kat Alexander, my book doula and thought midwife, who helped me via hours of conversation to spin my half-baked revelations into palatable (and hopefully relatable) paragraphs. David Blum, a teacher, mentor, friend, and the champion of my more foul ideas, pushed me to pay attention to what I pay attention to (no matter how gross it was). He also deserves a massive thanks for pressuring me to confront my hair shame and write "Bearded Lady," the piece that became the cornerstone of this book. My mom, Deena Altman, who is both my biggest fan and biggest critic, for poring through drafts with a nitpicky attitude that often made me cranky but always caused improvements, and even more so, for giving me a great shot at body acceptance by effortlessly modeling it since my childhood. Dad, don't worry, you get a thanks, too. Thanks, Dad! The bros—Matt and Logan, also! Dave Goldsmith, my husband and an extraordinarily private person, for allowing and trusting me to splash our life and interactions on paper for the world to see. Also, in a society that continually tries to convince us that we are perpetually flawed, I have been lucky to have a partner who reflects back to me nothing but wholeness (except for every other day when he tells me my pits could use a freshening up). And finally, I'd like to thank the fetuses developing inside me, the genesis of even more curiosity and awe. If a penis is growing inside me, does that mean I can say, "I have balls"?